THE GREAT PLAINS: A FIRE SURVEY

To the Last Smoke

SERIES BY STEPHEN J. PYNE

Volume 1, *Florida: A Fire Survey*

Volume 2, *California: A Fire Survey*

Volume 3, *The Northern Rockies: A Fire Survey*

Volume 4, *The Southwest: A Fire Survey*

Volume 5, *The Great Plains: A Fire Survey*

STEPHEN J. PYNE

THE GREAT PLAINS

A Fire Survey

THE UNIVERSITY OF
ARIZONA PRESS
TUCSON

The University of Arizona Press
www.uapress.arizona.edu

Printed in the United States of America
22 21 20 19 18 17 6 5 4 3 2 1

ISBN-13: 978-0-8165-3512-5 (paper)

Cover design by Leigh McDonald
Cover photo by Larry Schwarm

All photos are by the author unless otherwise noted.

Library of Congress Cataloging-in-Publication Data
Names: Pyne, Stephen J., 1949– | Pyne, Stephen J., 1949– To the last smoke ; v. 5.
Title: The Great Plains : a fire survey / Stephen J. Pyne.
Description: Tucson : The University of Arizona Press, 2017. | Series: To the last smoke / series by Stephen J. Pyne ; volume 5 | Includes bibliographical references and index.
Identifiers: LCCN 2016039131 | ISBN 9780816535125 (pbk. : alk. paper)
Subjects: LCSH: Wildfires—Great Plains—History. | Grassland fires—Great Plains—History.
Classification: LCC SD421.34.G74 P96 2017 | DDC 363.370978—dc23 LC record available at https://lccn.loc.gov/2016039131

♾ This paper meets the requirements of ANSI/NISO Z39.48-1992 (Permanence of Paper).

To Sonja,
old flame, eternal flame

CONTENTS

SERIES PREFACE

To the Last Smoke

WHEN I DETERMINED to write the fire history of America in recent times, I conceived the project in two voices. One was the narrative voice of a play-by-play announcer. *Between Two Fires: A Fire History of America, 1960–2012* would relate what happened, when, where, and to and by whom. Because of its scope it pivoted around ideas and institutions, and its major characters were fires or fire seasons. It viewed the American fire scene from the perspective of a surveillance satellite.

The other voice was that of a color commentator. I called it *To the Last Smoke*, and it would poke around in the pixels and polygons of particular practices, places, and persons. My original belief was that it would assume the form of an anthology of essays and would match the narrative play-by-play in bulk. But that didn't happen. Instead the essays proliferated and began to self-organize by regions.

I began with the major hearths of American fire, where a fire culture gave a distinctive hue to fire practices. That pointed to Florida, California, and the Northern Rockies, and to that oft-overlooked hearth around the Flint Hills of the Great Plains. I added the Southwest because that was the region I knew best. But there were stray essays that needed to be corralled into a volume, and there were all those relevant regions that needed at least token treatment. Some, like the Lake States and Northeast, no longer commanded the national scene as they once had, but their stories were interesting and needed recording or, like the Pacific Northwest or central oak woodlands, spoke to the evolution of fire's American century in a new way. I would include as many as possible into a grand suite of short books.

My original title now referred to that suite, not to a single volume, but I kept it because it seemed appropriate and because it resonated with my own relationship to fire. I began my career as a smokechaser on the North Rim of the Grand Canyon in 1967. That was the last year the National Park Service hewed to the 10 a.m. policy and we rookies were enjoined to stay with every fire until "the last smoke" was out. By the time the series appears, 50 years will have passed since that inaugural summer. I no longer fight fire; I long ago traded in my pulaski for a pencil. But I have continued to engage it with mind and heart, and this unique survey of regional pyrogeography is my way of staying with it to the end.

Funding for the project came from the U.S. Forest Service, Department of the Interior, and Joint Fire Science Program. I'm grateful to them all for their support. And of course, the University of Arizona Press deserves praise as well as thanks for seeing the resulting texts into print.

PREFACE TO VOLUME 5

IN THE SPRING OF 2012 I conducted two road tours of the Great Plains. One included a long trek around Texas with a jog into Oklahoma. The other swept through Nebraska and South Dakota. I revised and added essays on Illinois and Tallgrass Prairie Preserve that I had written previously. In August and September of 2014 I filled in two areas of interest that I had missed, Konza Prairie and the prairie potholes. To someone who had grown up in the Southwest and whose personal knowledge of fire resided in forests, those treks were a revelation. These essays are the outcome. Texas became its own novella—too large to include with the plains, too brief to stand on its own (sound familiar?), and perhaps too distinctive to sit with the other pieces. I decided to treat it as a regional suite in miniature.

Those who made my visit productive (and in some cases, possible) are acknowledged in the individual essays. But I offer my collective thanks again here. Special thanks to Mary Lata and Mark Kaib, who helped connect me to hosts and tutors. I also wish to thank Kerry Smith, who continues to edit the entire suite with care and insight.

THE GREAT PLAINS: A FIRE SURVEY

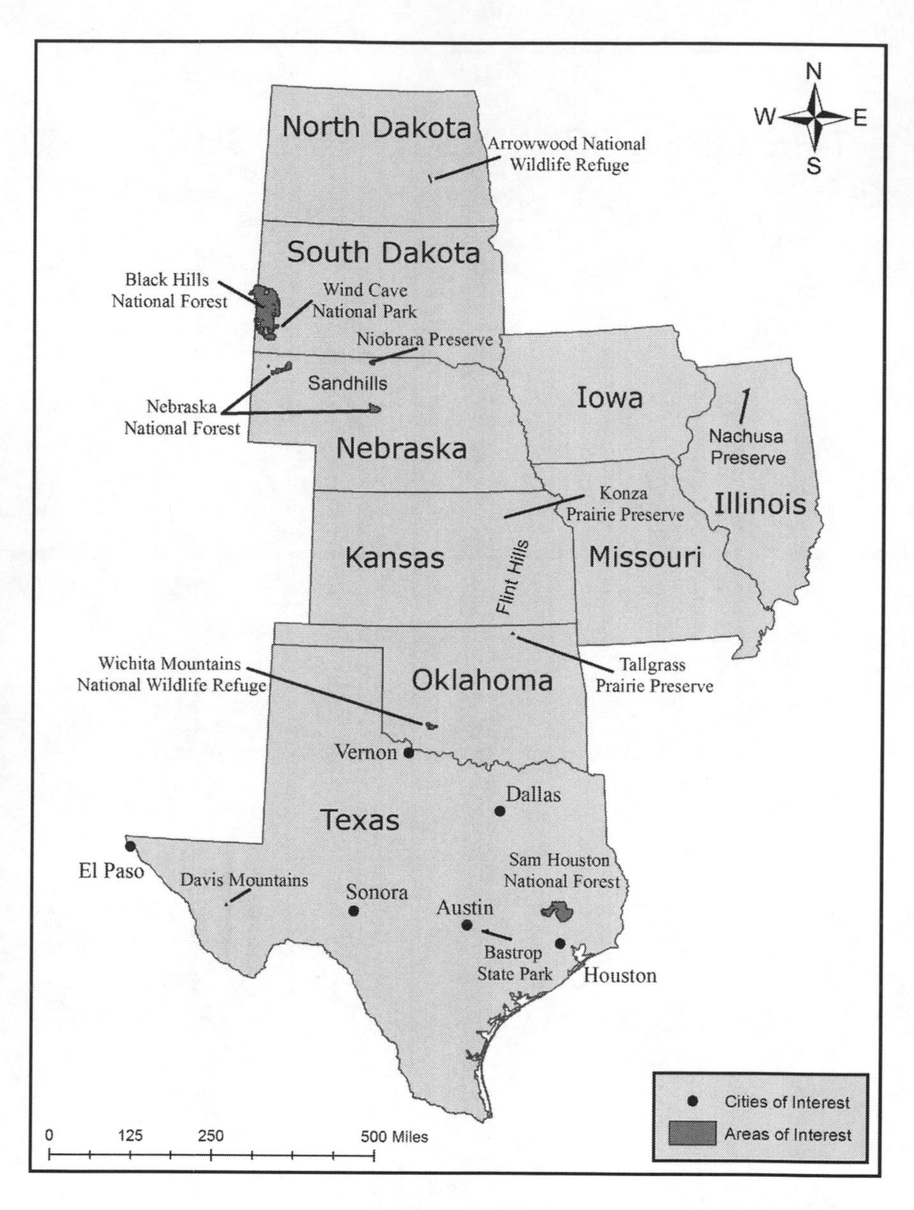

Map of the Great Plains

PROLOGUE

These Once-Conflagrated Prairies

From a belt of wood which borders the Kansas . . . we suddenly emerged on the prairies, which received us at the outset with some of their striking characteristics; for here and there rode an Indian, and but a few miles distant heavy clouds of smoke were rolling before the fire.

—JOHN CHARLES FREMONT ET AL., *REPORT OF THE EXPLORING EXPEDITION TO THE ROCKY MOUNTAINS* (1845)

As memory, as experience, those Plains are unforgettable; as history, they have the lurid explosiveness of a prairie fire, quickly dangerous, swiftly over.

—WALLACE STEGNER, *WOLF WILLOW* (1962)

WHEN EUROPEAN EXPLORERS reached the Great Plains, they found a terrain that was at once free roaming and featureless, in which they could see endless vistas but could often find no point on which to rest the eye. The effect could be disorientating, and for many, disturbing. They encountered grasses, thick and tall to the east, dwarfish and spare to the west. They found new creatures such as pronghorn, discovered vast herds of beasts such as bison that they had known only in small groups, saw familiar forests thin into stringers or clumps, and pondered indigenous peoples who wandered the scene seemingly in perpetual motion. And almost always they found fire.

Some fires burned in the spring before rains and green up. Others burned in the fall after prairie grasses had cured into stalks of kindling.

They burned amid summer droughts, and they burned through dry winters when snows banked into patches or melted away. Not for long was dry grass separated from spark, and never from winds, which blew, it seemed, forever, as implacable as gravity. For the plains, "prairie fire" became a kind of Homeric epithet like "wine-dark sea" or "shadowy mountains." By the 1830s random observations were yielding to paintings by footloose artists, reports by naturalists, and even pyric tourism. In his *A Tour on the Prairies* Washington Irving recorded how the Swiss count traveling with them wondered aloud whether their party might see a fire. Their guide assured them that if they did not, then "By Gar! I'll set one myself!"[1]

The "conflagrated prairies," as the Long Expedition called them, became an indelible part of settlement lore and literature, and later, science and politics. They merged with other natural hazards like frost, drought, and grasshoppers, and especially with those terrors brought by wind—with blizzards, dust storms, hail, and tornadoes. They were something that newcomers had to endure, a primal test of character. The prairie, as the saying went, "winnows out its own."[2]

Yet fire was different from those other threats. The vast majority were set by people; unlike thunderstorms or floods, they could not occur without the grasses that fed them. They could be set and attacked directly, and they could be contained by changing the prairie biota on which they depended. They would vanish with the wild landscape that sustained them. They were, in principle, something that could be eradicated like wolves, which made fire control a practical index of civilization. Primitive societies set fires and wandered around immense landscapes. Modern societies extinguished fires and lived in fixed abodes.

The Great Plains have long defied simple definitions. Look at them as a physiographic province, and they have one dimension. Look at them climatically, and they have another. And another if viewed ecologically. And yet another as history. In broad terms they are framed north and south, east and west, Pleistocene past and Anthropocene future, and woven by stories of American settlement; but that is a remote-sensed satellite-surveillance perspective. In detail, at the level of individual pixels and polygons, the nominal plains are pocked with hills, badlands, and escarpments; with anomalous patches of woods, prairies, and wetlands; with countervailing flows of movement that overspill historical levees. But nowhere are they far removed from fire.

Historically, they burned—they were the most regularly and widely burned province of North America. Their regimens of fire were traced by the shifting fortunes of grasses and trees, and by how settlement affected that border march. They were the first trans-Mississippi landscapes encountered by westering Americans, and among the last to be settled. Their narrative is loosely one of pioneering, but it is a frontier saga of recurring colonizations and collapses that breaks the continuity of the standard American narrative of settlement, yet it is the narrative that joins the region onto American experience, if sometimes obliquely.

Prairie fire entered into American art, literature, history, science, and politics. America's contribution to global fire art emerged from the 19th-century prairies. Prairies aflame found their way into settler accounts and chronicles, yielding only one of America's two national literatures on fire, matching that produced by the forested Great Lakes. It helped shape an American school of ecology. It factored into the politics and philosophy of environmentalism, from state-sponsored conservation during the Dust Bowl to private programs to restore tallgrass prairie. Yet, as Wallace Stegner suggests, the narrative seems to have happened in passing, like folk crossing the plains to get to Oregon, California, or Utah, and only later bonded to established narratives. The story of prairie fire is one that flared and then seemingly went out.

Yet fires did not wholly disappear. Like ecological insurgents, they retreated to hills that resisted the plow: the Flint Hills, the Black Hills, the Wichita Mountains, the Sandhills, the Loess Hills. They found niche habitats along railway rights-of-way and the occasional prairie memorial. They were herded into reserves, like the indigenes, left as patches of forest steppe, or passed by like watery potholes along the migratory paths of Canada geese and sandhill cranes. They flowed outward like spilled water when settlement stumbled. They revived when settlement stalled from drought and depression.

More recently, they have systematically returned, whether feared as feral fires or welcomed as a prodigal process without which restored prairie and ecological integrity are impossible. In part of the plains the conversion of grasses to farm continues—in North Dakota, aggressively. Elsewhere, a counterreclamation, as it were, is underway as people drain away from scattered farms and ranches and gather into towns, many removed from the region. Here aggressive grazing, farming, and tree planting are yielding to less deleterious practices, and with the lightened hand on the land comes more grass, and with grass comes the prospect for revived fire, and from relearned tolerance of fire has come enthusiasm for using it to renew landscapes degraded or lost by its absence. Where it had never left, fire's presence is strengthening. Where it had been swept aside, fire is returning.

The issue is whether it will return as conflagration or as a conservation practice, and how its deep boreholes into the geologic landscapes of fossil biomass animate the contemporary life that flows across its surface.

GEOGRAPHIES OF PLACE AND MIND

THE FIXED AND THE FLUID

Hard and Soft Geographies of the Plains

But more to be dreaded than this tribulation was the strange spell of sadness which the unbroken solitude cast upon the minds of some. . . . It is hard for the eye to wander from sky line to sky line, year in and year out, without finding a resting place!

—O. E. RÖLVAAG, *GIANTS IN THE EARTH*

The world is very large, the sky even larger, and you are very small. But also the world is flat, empty, nearly abstract, and in its flatness you are a challenging upright thing, as sudden as an exclamation mark, as enigmatic as a question mark.

—WALLACE STEGNER, *WOLF WILLOW*

THE GREAT PLAINS have proved hard to define, with as many definitions as there are criteria proposed. Even the sense of a stable core is disputed, one view envisioning it as a geographic center that binds the national periphery, and another as a spongy core defined by its exoskeletal fringe. The one imagines the plains as the national heartland; the other as heartwood whose role is to move materials to the cambial coasts that account for the national life. Either way the external frames tend to blur, which may be why so many accounts of plains experience focus instead on the self. It is hard for the mind's eye to find a defining reference at the center.

Yet it is possible to construct a rude matrix for understanding that begins with physical geography, segues into history, and concludes with some perhaps enduring features of life in the region. All translate into a characteristic pyrogeography.

Begin with the way the Great Plains divide along axes that run from north to south and from east to west. America's Great Divide lies not along the crest of the Rockies but along that wavy frontier between semi-humid and semiarid, for this is what broke the pattern of settlement.

The east-west axis is primary. Geologically, it describes a subcontinental plain that reaches from the Appalachians to the Rockies, roughed up here and there by hills that resemble stakes driven through a hide to hold it to the ground. Climatically, the eastern half is humid, and the western arid, with the divide between them roughly aligning with the 98th meridian, which separates rainfall capable of supporting dryland agriculture. This wavering line of rainfall traces an ecological boundary. To the east is forest, and to the west, grassland. Carl Sauer long ago noted that plains the world over that are subject to long fetches of wind tend toward grasses because, presumably, they can easily carry fire sufficient to retard forests. It is here, in its rough midsection, where the country is both level and grassy that America has traditionally sited the Great Plains.

The north-south axis conveys unity but not definition. The Great Plains have their southern terminus at the Gulf of Mexico, and their northern where grass yields to boreal forest or the chain of Great Lakes along the Canadian Shield. Climatologists may point to the peculiar sweep of the plains and its capacity to rapidly blast cold air south and warm air north in what has been termed a "climatic trumpet." The northern plains experienced glaciation, and still know winter snow. The southern plains avoided Pleistocene ice, and its biota can burn much of the year. These climatic blocs, north and south, hold most of the migration of fauna and peoples. Movement occurs within and around the two blocs, not across them.[1]

As with the nation, so with states within the greater plains. A few lie wholly east of the divide, a few west, while the core split the difference, some dramatically. For Kansas the difference is slight, a gradient of historical geography. In South Dakota the Missouri River runs close to that climatic track; the state severs into East and West River country. Texas tries, amid great strains, to span eastern woods and western desert under a single political system and narrative arc.

Beyond geography, history enters the quest for informing principles. Importantly the chronicle of American settlement in the plains mirrored its biogeography in that the primary passage went from east to west. However much a climatic trumpet might blow air, it did not move settlement. Cattle shuffled north from Texas, but failed; and some peoples moved south, but not many. Permanent occupation came from the east along trails that later became railways and highways.

The process of settlement, that is, ran cross-grained to North American geology; a folk migration spilled over the Appalachians, followed rivers and forests west until it struck the great prairies. There it stalled, regrouped, and then leaped over the plains to Oregon, California, and Utah. The plains were a thoroughfare that moved people and ideas from east to west. Only after Frederick Jackson Turner famously (or, as some would hold, fatuously) announced the end of the frontier in 1893 did settlement seriously encroach onto the grasslands; homesteading patents reached a climax 20 to 30 years after Turner announced that a line of settlement no longer existed. The Great Plains did not propose a separate story—did not serve to gather up and bind the divisions of the country into a national nucleus that could hold the country within its geographic lines of force. On the contrary, they offered them passage west.

The division of Civil War America into north and south carried across the Great Plains, politically with the Kansas-Nebraska Act and then by differing streams of colonizers. Almost all the settlement of the southern plains came from the South. Almost all the northern settlement came from the same European influx that swept over the North Woods. The two streams brought different assumptions about society, agriculture, and politics. In a 1923 essay about her native Nebraska, Willa Cather observed how New Englanders kept to themselves, "cautious and convinced of their own superiority," while those few "incomers from the South," although often kind, were "provincial and utterly without curiosity." "Colonies of European people," however, brought with them "something that this neutral new world needed even more than the immigrants needed land." While the Great Plains divided into humid east and arid west according to climate, they separated north from south according to cultural heritage.[2]

The expansive prairies posed many puzzles to westering Americans. What caused those horizonless grasslands? Even Thomas Jefferson and John Adams debated the two competing explanations (climate versus fire), with Jefferson favoring fire at the hands of the indigenes. What did it mean for the prospects for settlement? Aridity, grass, and terrain all made separate appeals. Expansive grasslands were a problem because they meant a scarcity of wood. Terrain—or lack of it—was tricky because it disoriented and destabilized the traditional siting of settlement into valleys. But aridity posed perhaps the greatest conundrum because it undermined notions of water—its use, laws, and politics. Zebulon Pike, journeying to the Southwest as Lewis and Clark did to the Northwest, thought the western plains a "desert." The 1819 to 1821 Long Expedition formalized that expression into a Great American Desert. In 1878 John Wesley Powell argued that western aridity demanded a fundamental reconstruction of homesteading practices and the need for government intervention to prevent abuse based on water monopolies. In 1931 Walter Prescott Webb used grass, aridity, and terrain to inform his masterwork, *The Great Plains*, which did for the region what Turner did for the frontier and Powell's classic did for the arid West.

The plains were where all those elements converged. Half the plains aren't arid; half did not form a historic line of settlement; half rumple into badlands, hills, and sumps. But what the country came to regard as the Great Plains proper was where terrain, climate, and flora passed through a phase change from east to west, and where the national epic of pioneering broke away from its inherited plotline. The plains challenged the old technologies, ideas, and institutions that had sustained folk colonization. They scrambled the westering frontier into patchy narratives of mining, trapping, herding, logging, and farming. Not least, they divided the bulk of private lands, which reside east of the 100th meridian, from public lands, which mostly lie to the west. In brief, the Great Plains separate the East, broadly conceived, from the West.

The plains ruptured the continuity of American westering, and surely this helps account for the omission of the plains as a separate plot in the national story of fire. Their role has been to trace a divide between other regions, not to furnish an integral history of their own. Many early explorers believed the American republic would cease to expand when it struck the plains amid a setting so different from what had sustained the trans-Appalachian frontier. Instead, they became a place in which narrative, like pioneers, passed through. Their contribution to the national culture of fire has been remarkably rich and just as easily overlooked. The Great Plains blew through the national narrative with the flash and brevity of a prairie fire.

Among their geography of dichotomies lies another, and it may be the most fundamental. It organizes the critical features of the plains into a double matrix, a hard and a soft geography, that distinguishes between the fixed and the fluid.

The fixed are those features that reach into the earth. Mountains with deep geologic cores, plants with deep roots or dense subsurface masses, buried water and fossil biomass. The fluid are those processes that sweep over the surface—water, wind, dust, migration, fire. The fixed underlie and hold the land; beneath Kansas lies a triple junction where tectonic plates once met to define the cratons of North America. The fluid pass through or across that landscape. The climatic trumpet has analogues for flora, fauna, and people who have come and gone, filled and emptied the region with quickening pace throughout the Pleistocene and beyond. One part of the plains endures. The other blows and passes. That slice of their natural history that deals with the American era might aptly be explained as an attempt to fix what is fluid by joining the surficial to the deep, to give roots to what is, by nature, inclined to move.

The clash became doubly striking when the plains encountered a people with similar tendencies. As one French critic has written, the Cartesian premise of American society is, "I move. Therefore, I am." Yet a society needs to anchor itself, and from the early discoveries, how to plant such a society to such a landscape perplexed observers. The Great American Desert as an idea became as peripatetic as the Seven Cities of Cibola and as elusive as the Great American Novel, but its origin in the plains began with Long's observations regarding "sandy wastes and thirsty inhospitable steppes." Little water, little wood, few navigable streams—the western plains were an "unfit residence for any but a nomad population." Their likely destiny was "for ever" to "remain the unmolested haunt of the native hunter, the bison, and the jackall." They seemingly defied the kind of rooted settlement even footloose Americans saw as the legitimate outcome of nation building. They would remain a frontier marchland.[3]

The debate over the fundamental character of the Great Plains as an abode for human habitation thus began early. It's worth remembering that "desert" was etymologically related to "deserted," and that when Robinson Crusoe spoke of a "desert isle," he referred to a place devoid of people. A desert was a place that could not support stable populations except at oases or that had once known inhabitants who were now gone. The earliest American explorers thought the western plains unsuited for settlement, that away from rivers it would always be a landscape of migrants. The first American occupants were in fact explorers, trappers, emigrants on their way elsewhere, and drovers. When farming tried to take root, it needed substitutes for surface wood and water. It turned to barbed wire and coal for the one, and to windmills for the other; and it planted trees. It tried to fix life on the surface by connecting—rooting—to the subsurface.

As geologists learned more about the Pleistocene history of the plains, they saw a recurring pattern. As the land swung into and out of glacial epochs, the landscape filled and emptied with creatures on an immense scale. The land always seemed to be out of sync. The solution was to move—to migrate seasonally, to trek into and out of the region, to move into and out of history.

Paleontologists and anthropologists have found analogous events in the Holocene; and historians have traced similar chronicles in Anthropocene America not only from climatic wobbles, including ferocious droughts, but from economic cycles. Drought and depression became the primal movers of life. They set the rhythms of overpopulation and depopulation. When they combined in the 1930s, the Great Plains reentered the national narrative and put the region once more on the move. In recent years chronic recession has renewed the trend to the point that critics have revived the concept, in a more benign avatar, of a Great American Desert.

If you were wary of mobility, the solution was to put down roots. Tapping down into water, oil, and gas helped fix fluidity, but at some point those sources would be tapped out, and the fluid would reassert itself over the fixed. The surface would have to exist on its own terms, not through drill holes that staked it to the legacy of time. When that happened, the Great Plains would default to that long-defining feature at which sky and wind met earth: their grasses. "The history of the Plains," Walter Prescott Webb wrote, "is a history of the grasslands." Even farming sought to replace wild grasses with domestic varieties.[4]

Their grasses are critical, too, because they are fundamental to an even more powerful index of life on the plains. They are the arena and fuel for free-burning fire.

GRASS

A Pyric Primer

IN 1929 J. EVETTS HALEY offered a panegyric to Great Plains grass. It was "essential to every form of animal life." It "alone" accounted for the fabulous herds that had roamed the landscape, and then for the livestock that replaced them. A number of plains creatures could survive without water, metabolizing it from forage; none could live without grass. Where grass was good and water poor, settlers sank wells and windmills, and survived. Where there was adequate water but feeble grass, they failed. Every patch of "good grass land" was in productive use; "much other land is still unappropriated" to any purpose. He concluded, with unanswerable logic, that "its preservation was of the utmost importance."[1]

Grass. It's what defines the plains in popular imagination. Wild or domesticated, it's the basis for plains economy. It's what burns. It's what appears to simplify an understanding of fire on the plains and what complicates its biology, for the grasses are many, their diversity far greater than America's woods, their biotic communities kaleidoscopic, and their relationship to fire vastly more nuanced than the usual dichotomy between grass and tree pretends. Grasses make a simple fuel, easy to model for fire behavior, but a complex ecology, often baffling to oversee. Plains grasses feed every form of plains combustion, fast and slow. They were a matched set, an ecological tag team. The story of fire on the plains is the saga of the rise, fall, replacement, renewal, and survival of its grasses.

GRASS AS FUEL AND FODDER

Although used interchangeably in common speech, grasslands and prairies differ. Grasslands, strictly speaking, have grass and little else. Prairies include rich medleys of grasses, forbs, and sedges, and are often salted with woody species. The Great Plains have both.

In the usual typology prairies segregate into tall, mixed, and short along a climatic gradient. The tallgrass dominates in the semihumid east, famously pushing into the Ohio Valley as a lush "prairie peninsula." The shortgrass is its semiarid western counterpart, lapping to the Rockies. Where the mountains flex westward, so does the shortgrass steppe, making a shortgrass prairie peninsula in Montana to echo the tallgrass one in Illinois. The mixed grass, as its name implies, negotiates the mobile and mingled border between them. The lusher patches can hold 150 to 300 species. Those landscapes also include trees—conifers to the north, south, and west, and hardwoods along the eastern midsection, with invasive juniper scattered throughout.[2]

Whatever their particular composition, the grasslands are never far from fire. In fact, since the Miocene, grasses have reconstituted fire regimes across the Earth, and with the arrival of C_4 grasses in warmer climes, they have remade whole landscapes in their own image. The reason is simple: small-particled and responding quickly to changes in humidity and heat, grasses make an ideal fuel. They can burn year after year. They propagate flames that can wipe out the seedlings of slower growing, larger competitors such as woody shrubs and trees. In truth, many "forests" are savannas or grassy woodlands whose fire regimes follow the seasonal rhythms and spread of grasses. Those fine fuels create a positive feedback. So long as fire persists, the grassland will inform the ecology of the landscape.

That, at least, is how modern fire behavior imagines grasses, with flame like an oil drop suspended between the charged plates of climate and fuel. What is actually available to burn, however, depends on the competing slow-combustion of metabolizing prairie fauna, which consumes grass and forbs according to their own rhythms and migratory habits. The effect is obvious with the large herbivores—the bison, the deer, the elk, the pronghorn, the classic menagerie of big-game prairie creatures, later replaced by a monofaunal culture of cattle. But that consumption is likely rivaled by small mammals such as rabbits, voles, mice, and of course the prairie dog, whose communities can be vast. Even more astonishing is the contribution of insects. Herbivorous grasshoppers and butterflies, and granivorous beetles, ants, thrips, and other species, feed on grasses and forbs, and help structure their ecology. It is estimated that, at any one spot, the leaf-eating insects could consume as much as bison. The regimen of fire does not follow from simple flame-fuel dynamics since fire can only burn what fauna have not. In fact, some ecologists have even likened fire itself to another herbivore.[3]

This swirl of consumers creates a complex choreography of combustion, what has been characterized as fire ecology's three-body problem. Take the simple case of bison. They prefer grass over forbs and overwhelmingly feed on grass regrowing from recent burns; grass stalks two years old are ignored as essentially inedible. What remains of that old, uneaten range, however, feeds fire. Such patch-burn dynamics are tricky enough to predict, but throw in the ensemble of other grazers, and then the insects and voles and prairie dogs, and the predators, from wolves and grizzlies to mantids that shift the populations of those grazers, and then add climatic variability, and the resulting fire regime becomes anything but stable or simple. Compared to multilayered fuel arrays in forests, grasslands are easy to model, and to burn. But factor in the biological dynamics that fashion those grasslands, and fire regimes quickly leap into ecological hyperspace. The prairies, as Karen Smith has noted, are a "botanist's paradise and a manager's nightmare." Reducing the grasslands to fuel particles is of a piece with treating them solely as forage for cattle. It simplifies practice. It complicates understanding.[4]

So long as dry grass exists, the land can burn—spring, summer, fall, even winter when snows melt off. Not much grass needs to remain on the ground to carry a skimming flame in high wind. Burning appeared expansive because it was as chronic and unrelenting as the wind.

For early European explorers the association of fire and prairies was a given. Where you had one, you had the other, and precocious tourists like Washington Irving expected to experience a sea of flame as much as swarms of bison, and his French Canadian guide was happy to ensure they would get one.

That casual comment points both to the role of people as fire creatures amid the plains and to the confusion over how to incorporate them into theory and story. Clearly people competed with plants and animals; they hunted bison and tended sunflowers; and, in a hundred ways, they helped restructure the biota, which is to say the capacity of the landscape to burn, and more, unlike any other creature on the scene, they actually lit fires. They have done it since those grasslands began to form at the end of the Pleistocene. If the expansiveness of barrens and of west-increasing grasslands in the form of prairie patches and peninsulas that spread into vast steppes both intrigued and baffled early Europeans, so did the role of fire, and of people as fire-starters.

From the beginning a debate has flourished between those who have sought an explanation in soils and climate and those who have thought the answer lies in fauna and fire. Were those fires informative, without which the prairies could not exist? Or were they an epiphenomenon, something that flickered across the surface like summer heat lightning, something that happened often on prairies but was not critical to them? And if grass made the fires possible, might it also be true that fires made the grasses? Deciding whether fire merely associated with grasslands or whether it created them has mattered to grassland science and even more to management. One group favors a wholly physical explanation. Prairies are a product of climate and soil, with soil a derived property of climate, and fire an aftershock of climate-caused grasses and lightning. In this conception the grassy combustibles power fires that hurl against heat-resisting trunks of trees, and their relative strengths determine whether grass or tree prevails. The competing view appeals to a more biological explanation. There is little tall about a tallgrass prairie that is grazed, and those sites are most grazed which grow where they have recently burned, since they are more accessible and far richer in protein. The fast combustion of flame has to compete with the slow combustion of metabolizing bison, elk, pronghorn, prairie dogs, and grasshoppers. In this scenario prairie patches obey a dynamic set by biological agents, within which the burning is embedded. In principle, this makes sense since fire, while not alive, is a creation of the living world. In practice, it means that an intricate choreography of burning and grazing shapes the landscape.

But the deeper debate is about people. Thomas Jefferson discussed the issue with John Adams and concluded that the practice of "fire hunting" by the indigenes was the likely source for the prairies' fires, and the fires were "the most probable cause of the origin and extension of the vast prairies in the western country." Aldo Leopold finessed the question of fire sources in favor of fire sinks—settlement's barriers to fire's spread, which led to thickening woodlands. "In the 1840's a new animal, the settler, intervened in the prairie battle. He didn't mean to, he just plowed enough fields to deprive the prairie of its immemorial ally: fire. Seedling oaks forthwith romped over the grasslands in legions, and what had been the prairie region became a region of woodlot farms." Carl Sauer frankly concluded that the burning was everywhere that could burn and was anthropogenic: "Wherever primitive man has had the opportunity to turn fire loose on a land, he seems to have done so, from time immemorial; it is only civilized societies that have undertaken to stop fires." Like many observers he noted that the proneness of landscapes to propagate fires splintered to the eastward, as land roughened, watercourses multiplied, and humidity thickened. As much as anything, the human torch made the Great Plains the kind of grassland they became and then maintained them even in defiance of climatic pressures that would, unchecked, have propelled them into woods.[5]

Still it seems implausible to some commentators, and undesirable to others, that aboriginal humanity, outfitted with spears and torches, could have prompted such immense effects. One reply is to note the difference between creating such landscapes and maintaining them. Surely, too, fire—that most interactive of biotechnologies—worked in close coupling with other factors. Moreover, the 100th meridian that divides the humid from the arid also segregates sources of ignitions. There is ample documentation of lightning starting fires in the western grasslands—not as many as the record of fires chronicles and ecology demands, but a permanent presence nonetheless. To the east, not so. The storms are too wet to allow fires to kindle, and the landscapes too dissected with streams and wetlands to allow them to spread. The fires that the tallgrass prairie needed to flourish could only have come from people.

This introduces another complication for understanding and management. Natural science dislikes including people because it complicates the modeling and forces hard "drivers" like quantifiable climate to compete with sloppy cultural motives. Preservationists dislike including people because it means nominally pristine nature must continue to include them. If the prairies are invented landscapes, then their future depends on people reinventing them. That complicates the politics. It muddies the ethics. It could seem to sanction an obstreperous humanity to persist in its arrogant meddling when the future will depend on staying humanity's hands, not urging them on. The situation recalls the comments of a 19th-century critic of Darwinian evolution: that we should hope it isn't true, but if it is, that the word doesn't get out.

So it matters how people lived on the plains and how that living evolved over time. That they burned methodically (and accidentally) can hardly be doubted. (If they had not, they would be the only peoples on the planet to deny themselves that power, and they would be living in woodlands, not grasslands.) First-contact accounts, from Cabeza de Vaca to Father Louis Hennepin, testify to indigenous burning to assist hunting. His survey of 17th-century exploration literature led Carl Sauer to conclude that fires were set everywhere that would support them. But far from lighting and letting fires rip hither and yon, native burners operated within a rough system. Their fires were constrained by the available fuels, which depended on the peculiar dynamics of grazing, which in turn followed in good part from past burning. Only in a few places and at exceptional times could fires burn untrammeled for long. In tallgrass there were frequent natural barriers as well as an ecological quilt of patches variously available to burn. In shortgrass the fuel load was lighter overall and more easily disrupted by herds and prior burns.

The fact is, fire was part of a human economy of land use, itself embedded within extraordinarily fire-prone landscapes but in places full of buffers and baffles in which lightning competed with anthropogenic ignitions, and fauna with flame.

THEN . . .

What thickens the plot, however, is not only that humans moved around their home landscapes, but that they came and went as tribes. Trade and diseases in advance of white settlement broke and reassembled relationships among peoples. The propagation of the horse revolutionized the scene: it gave existing tendencies toward mobility even greater force. The frontier of American settlement, often accompanied by forced relocations, pushed tribes west. The Five Civilized Tribes were uprooted and sent to Oklahoma. Some classic "plains" tribes such as the Sioux had only moved into the grasslands a century or so before. For almost the entire tenure of European exploration and subsequent colonization, the peoples of the Great Plains had been in upheaval. That turmoil ramified throughout the region.

When early settlers reached the unplowed prairies and began documenting fires, they were not recording the natural scene, or even a traditional one, but likely a fire regime that was unraveling and unhinged. The top grazers, the bison, were being extirpated. The alpha creatures of fire were killed off, pushed into migrations, or sequestered onto reservations. Between the collapse of the old biota and its repopulation lay an ecological vacuum and a historical lacuna that was ideal for expanding the realm of free-burning fire. The old regimes had been heavily eaten, broken into incombustible patches around prairie dog towns and buffalo wallows and poorly combusting swathes where herds of bison and hordes of grasshoppers had grazed. With relentless wind, long stretches might burn, but spottily, here feasting on a patch of old grass, there skipping over swales mown by prairie dogs and pronghorn.

The sweeping prairie fires of settlement may have been an artifact of the settlement process itself. They were not a confrontation with untouched nature, but an encounter with a feral setting, many of whose ecological checks and balances were scrambled or blurred. For 15 to 20 years after the Civil War, combustibles had flourished outside their usual regimens of grazing and burning. Wildfire spread through the landscape like rats in a plague city. In 1832, in what is the first text-and-image report on prairie fires, George Catlin painted two scenes. In one, two deer flit over light flames—most likely a fire creeping through a grazed setting. In the other, horses and riders race at full gallop ahead of a rolling-thunder fire front; but Catlin notes that those lush fuels grew in river bottoms. What happened in the early rush of settlement was that the benign flames went out and the violent ones propagated, like cedars at a later date, from their restricted niches to spread over the countryside.

Those fire blizzards had to be countered if settlement was to pass from the wild to the civilized. If the prairie fire was a constant in explorers' accounts, the fight against it is one of the set pieces of plains pioneering. In the north it involved communities, which rallied against a common foe, turning out with plow teams, burlap sacks, shovels, and torches. In the south the fight turned on large landowners and their ranch hands, who liked the task no better than sodbusters did, even if they replaced burlap with the beef drag in which they split a steer from head to tail, roped hoofs to horses, and dragged the carcass across the flaming front. As Haley noted, there was "no harder, hotter, or more trying work done on the western ranges than fighting burning grass." Worse, to lose the winter range threatened the whole economics of open-range livestock. The best compromise was to plow "fire guards" around any site you wanted protected, preferably double or triple lines with the medians burned out. Haley affirmed that cowboys liked this even less.[6]

These were temporary measures. A full solution demanded converting the wild grasslands into something else. For farmers this meant plowing fire under, and for ranchers it meant grazing it away by stocking as high as possible, even overstocking. Where the strategy—if it can be called such—worked, it temporarily squelched fire. But keeping fire out meant the land kept changing, and the long-term consequences pushed the land towards woods, which diminished pasture and imposed its own fire regime.

What makes the settlement scenario particularly fascinating is that it has been repeated, flickering with gusts of population. In the first wave, fire blew over the land like a gas released into a vacuum because bison and indigenes were gone, and cattle and newcomers had not built up numbers to compensate. That took decades, during which wildfire continued to plague the plains. Locally, there were still gaps and extensive landscapes neither wild nor cultivated. The 20th century was well advanced, as Julie Courtwright notes, before "even 30 percent of the plains felt the plow." Hamilton County in Kansas remained open range "until 1931."[7]

The suppression of wildfire by means of systematic settlement was the assumed strategy for the Great Plains, as it was for pioneering in the North Woods, and it is not news that, until colonization had converted the land, fires continued, like outbreaks of disease among populations as yet unvaccinated. What threatens to jar that narrative out of its familiar ruts is the reemergence of fire in second-generation settlements. Land was claimed yet not fully cultivated, and pasture might become as rank as prairie. Fire returned, often as "the worst" on record. The moral was, only continued maintenance and vigilance could keep wildfire on a short leash.[8]

Plains history, however, has never shown much stability for long. Soon after settlers filled up the plains, they began leaving; population had once again overshot. The Dust Bowl inspired at least locally significant exoduses. On marginal farmlands economic incentives led youngsters to move out. When the pace quickened toward the end of the 20th century, alarms sounded over a new Great American Desert in the making, and a proposal emerged to midwife with deliberation what was birthing by accident and so convert large swathes into something akin to their presettlement state, what was termed a Buffalo Commons. If anything like such a conception occurred, it would return fire along with bison, and probably in a more virulent form since the full complement of fauna would be absent. The Great Plains would face the choice that other wildlands, legal or de facto, do: they could have wild fire or controlled fire. But so long as they remained in grass amid a climate at least occasionally, if not seasonally, arid, they would have fire.

. . . AND NOW

Today, an estimated 96 to 98 percent of the original tallgrass prairie is gone, and what remains is mostly clustered around the Flint Hills of Kansas and Oklahoma. Interestingly, this is about the same survival rate as for longleaf pine, and like the pine, the tallgrass is an important fulcrum for environmentalist or restoration causes, and one that argues for an interventionist hand. In part this is a matter of scale. Most prairie patches are too small to retain a migratory grazing regimen anything like what they knew prior to settlement. And in part human ignition is needed because lightning will never start enough fires to favor grass over woods. It's hard to resist the argument that people were fundamental in establishing the post-Pleistocene prairie; it's harder to argue that modern remnants or reconstructions will happen without them. These are, though in environmentalist garb, working landscapes. Without humanity's busy hands the prairies will soon sink under the weight of invading trees.

That the prairies were maintained by burning was appreciated early by thoughtful observers. When fires stopped because of grazing or the removal of the torch, trees soon seeded in and remade the scene. Pioneers reported the process as they moved west—at Big Lick, Kentucky, across the prairie peninsula through Illinois, over the hills of Missouri. Thoughtful observers actually tracked the progress of settlement by the height and density of the woods that sprang up in their wake. Mostly, in the well-watered east, the trees were accepted, even welcomed. In the Great Plains proper the process was assisted by deliberate planting. More recently, however, shrubs and trees have become woody weeds that threaten to overwhelm grasslands and convert the region into overgrown woodlots. In the old days people thought they could eliminate fire indirectly by the simple act of settling. In this second-order variant, a new pattern of settlement (or unsettlement), attuned to a modern economy, has revived the process, this time with volunteer species such as juniper. Eastern red cedar, in particular, serves as a pioneer intent on replacing grass, not this time by plowing it under but by overgrowing it. Having determined what portions of the plains need to remain in grass, landowners are now finding that woods are overrunning them. The insurgent trees are destroying ranching livelihood as surely as the loss of bison did indigenous hunting in the late 19th century.

It can be a hard sell to argue that what the land needs is what a century and a half of onerous pioneering struggled to take out. The fear of fire lingers, its horrors burned into the neurons of old-timers, its lessons chiseled into the retold legends of pioneering. Apart from psychological issues, however, no reinstatement is possible without fuel; and in range, this means a loss of forage, which is to say, of immediate income. In forests the "fuel problem" is having too much, and officials look to prescribed fire to reduce those combustibles and to give the land a much-needed ecological jolt. In grasslands the fuel problem is too little, and fuel management means building grass reserves. Taking land out of production can be as costly as paying harvesters to thin woods. The reason, as with so much of the prairie, is that the relationship is not simply between fire and fuel, but between fire, fuel, and fauna.

Whether for commercial production or ecological integrity, Haley's and Webb's observations still hold. For the Great Plains their grasses are their essence. In Haley's day protecting grasses meant keeping fire out. Today it means putting fire in. It's a shift in the climate of opinion as radical as that by which Earth switched in and out of glacial epochs. But without fire, its "immemorial ally," as Leopold called it, the prairie will choke, starve, shade out, and fade away.

SEASONS OF BURNING

Prairie Fire in American Culture

THE GREAT PLAINS lie within the center of the country and close to the historic core of its pyrogeography, yet today they reside on the periphery of a national fire culture and its accompanying narrative. The paradox is even more peculiar because the plains gave the nation its only school of fire art and one of its two major fire literatures, educated foundational figures in what evolved into fire ecology, and became an exemplar in the debate over private and public conservation. The easiest explanation is that prairie fire, like the western plains, was met and then leaped over, and by the time the country reengaged, those fires were receding and the country's major institutions set. Even as the Great Plains were being pioneered, the country had moved on.

How the region went, culturally speaking, from fiery core to burned-out cinder has its curiosity as a study in historical contingency. It was largely a matter of timing. It seems that in history and culture, as well as in nature, season of burning matters.

IN ART

Some countries create a fire art, some do not. It helps if fire is a part of quotidian life or fuses with its sense of the national sublime, but fire by itself does not make art; art does. Russia produced a Ural school of forest fire painting in the late 19th century. Canadians, with a similar boreal landscape, painted nothing. Australians have an unbroken legacy of fire painting and poetry. South Africa has almost none. For art to emerge it has to speak to cultural interests and have on hand suitable genres to express them. America's flaming prairies were ideally situated for both.

Painting was morphing from the Grand Manner—large canvases that illustrated enduring moral lessons—to landscape in which natural history replaced human history as a source of inspiration and emulation. America was especially receptive to this shift since it had no monuments from antiquity, no coliseums or Parthenons, but it had an abundance of wild nature. It had Yosemites, Yellowstones, and Niagara Falls, and later, Grand Canyons. None, however, captured the congested drama of battles that served so often as the backdrop for the Grand Manner. Prairie fire did. The flames divided the field into competing realms: they forced action and they drove herds and people. For artists increasingly fascinated with light, as the 19th century was, fire offered a glorious palette.

An art drawn to nature in the wild met an expanding frontier aflame. Artists joined other explorers and journalists in reporting on the national estate and the saga of pioneering that was claiming it. But they could also make a reputation as an artist from such scenes; and many did, or having painted scenes from the American West generally, they added the burning plains. Once established, the genre perpetuated itself until the prairie fire was no longer novel and, with the advent of modernism, art found art itself more interesting than the things it had formerly represented.

The process began with George Catlin, who painted several famous scenes from his 1831 voyage along the Missouri River, each of which established a trope. One showed roiling smoke and flames along the horizon, with a Native American family watching from the foreground in nervous alarm. Another is a close-up of rolling prairie with the fiery front wending like a rivulet of flame: "Where the grass is short, the fire creeps with a flame so feeble that one can easily step over it. The wild animals often rest in their lairs until the flames touch their noses. Then they reluctantly rise, leap over the fire, and trot off among the cinders, where the fire has left the ground as black as jet." Such fires, Catlin notes, are "frequently done for the purpose of getting a fresh crop of grass for grazing, also to secure easier traveling," but they have an aesthetic no less than utilitarian outcome: "These scenes at night are indescribably beautiful, when the flames, seen from miles distant, appear to be sparkling and brilliant chains of liquid fire hanging in graceful festoons from the skies, for the hills are entirely obscured."

A third trope moves the action to the center as a mounted band flees before an approaching flood tide of monstrous flame. Catlin jotted breathlessly into his journal that

> there is yet another character of burning prairies . . . the war, or hell of fires! where the grass is seven or eight feet high . . . and the flames are driven forward by the hurricanes, which often sweep over the vast prairies of this denuded country. There are many of these meadows on the Missouri, and the Platte, and the Arkansas, of many miles in breadth, which are perfectly level, with a waving grass, so high, that we are obliged to stand erect in our stirrups, in order to look over its waving tops, as we are riding through it. The fire in these, before such a wind, travels at an immense and frightful rate, and often destroys, on their fleetest horses, parties of Indians, who are so unlucky as to be overtaken by it.

The two polarities Catlin painted established the genre. That hellish onrushing fire, with the rolling smoke like a thunder cloud and the flashes of flame like lightning, became the dominant design as artist after artist painted surfs of flame that drove bison, elk, antelope, wolves, sheep, cattle, and people before it. William Hays split the scene in half with a roiling fire driving vast herds before it from left to right, as though the canvas were a map. Gustave Doré had panicked sheep rushing toward the viewer. A variant imagined the scene of a small band under threat. Charles Deas painted such flames bearing down on a mounted trapper and again threatening a wagon train. Alfred Jacob Miller showed trappers and indigenes responding by setting backfires and swatting out the fires on the downwind flank. Nathaniel Currier and James Ives dramatized for their popular lithograph series the more common response in which the party kindles an escape fire into whose rapidly burned patch they would move to ride out the fast-encroaching front (see A. F. Tait's *Life on the Prairie: The Trappers Defense, Fire Fights Fire*).

Similar artwork appeared well into the 20th century even from the hands of the country's most celebrated painters and illustrators. Frederic Remington painted Indians setting grass fires and ranch hands quenching them with "beef drags," and desperate cowboys driving herds ahead of the consuming flames. Charlie Russell showed the Crow tribe burning the Blackfeet range; bison and antelope crossing a river to escape stampeding flames; and assorted campfires, signal fires, and cooking fires. The larger list goes on to include locomotives belching sparks while wild flames and bison race alongside, whole galleries of fire-driven stampedes, and many, many copies and colorizations of the classic images.

More recently, the revival of enthusiasm for environmental matters—and hence for burning prairie—has rekindled the interest of contemporary artists, although, like most art, their deepest instincts are to imitate the masters or to work within the genres of their time. That, after all, is what makes it art.[1]

IN LITERATURE

In the same way that art comes from art, literature comes from literature. The pivotal figure seems to be James Fenimore Cooper and his Leatherstocking Tales. The first, *The Pioneers* (1824), opens in upstate New York and ends with Natty Bumppo and Chingachgook trapped in a forest fire. The sequence concludes geographically with *The Prairie* (1827), in which the Leatherstocking, now identified as the Old Trapper, guides a group onto the prairies. Along the way the sagacious frontiersman saves his party by kindling an escape fire. The push of American settlement, originating in the northern woods, concludes when it strikes the grassy plains. The torch passed from Adirondacks to Great Plains.

The Leatherstocking Tales are widely recognized as a kind of creation story for the American experience, the beginning of its fascination with the frontier. Cooper was himself the Old Trapper of the western novel. Thereafter literary men traveled to the plains for the same reasons artists did. Washington Irving followed with his *A Tour on the Prairies* (1835); explorers, travelers, and later newspapermen added to the roster. An increasingly literate populace, or at least one that carried a tradition of belles lettres among the baggage, created a popular body of written work on fire. With the prairie fire the expression "illuminated manuscript" took on new connotations.

Cooper's oeuvre foreshadows the two regional literatures of American fire and the way they are linked through the flow of settlement. The Lake States have a literature of forest fires overrunning newcomers, a grand narrative that ends with the land cleared and the fires removed. The Great Plains have a comparable tale of homesteaders and small towns resisting or being overtaken by grassfires that likewise concludes with the flames banished by plowing, roading, gardening, grazing, and replacing. (Even classics such as *Little House on the Prairie* have an obligatory set-piece fire.) For both, the blowups of the past help measure the heroism of the pioneer. But prairie fire has become a story from the vanished past, along with cholera epidemics and locusts. The rootedness of settlement had no place for something as free flowing as wind-driven flame. Wildfire went from being a recurring theme to a literary trope.

What is striking is that there is no similar literature from the southern plains or, for that matter, from the southern pineries. Settlement proceeded at an earlier time and from a culture less bound to formal art and literature. While settlers in the north were recording prairie fires within a narrative of endurance against the elemental plains, the classic accounts from the south were about droving, and although wildfire was a threat, the fires that populate accounts like Andy Adams's were campfires. The southern narrative is about moving: prairie fire is, as it were, just another variant of the plains' tendency toward the fluid. The northern narrative is about staying put, which makes fires a threat. Like its authors, the literature was rooted in a place.

Outside the Great Plains, however, or for that matter outside the Great Lakes, it did not spill over into a distinctively American genre, any more than the art of prairie fires did. Instead the grand narrative of American fire turned on the creation of a permanent public estate—these were the landscapes that mattered to the nation overall. The Great Lakes conflagration is replaced by Great Fires and Big Blowups. Eventually, a culture of fire emerged out of that experience, passed from generation to generation, and much later a literature emerged, though one made de novo and curiously divorced from its antecedents.

It would not be far off the mark to date the reemergence of such a literature to 1992 when Norman Maclean posthumously published *Young Men and Fire*. That book did for contemporary fire, grounded in the public lands of the Far West, what Cooper's Leatherstocking Tales did for the trans-Appalachian frontier. The central action of Maclean's meditation turns on an escape fire lit by foreman Wagner Dodge, which Maclean regarded as an unprecedented invention. It took a colleague to remind him that Cooper depicts just such a fire in *The Prairie*; but there was no continuity between the techniques frontiersmen had long used in tallgrass prairie and Dodge's daring act amid the piney savanna of Mann Gulch, and no literary continuity apparent even to a professor of literature such as Maclean. The written legacy had broken, had been left with the sod house and the mule-drawn plow. The prairie fire had passed from the national narrative and a regional variant would have to be reinvented.

IN SCIENCE

With its fractal patches and edges America's unsettled rural countryside was ideal for wildlife and inspired many a youth to become a naturalist. Not a few of those who figure in postwar environmentalism (think Aldo Leopold) and in the fire revolution (see Herbert Stoddard and E. V. Komarek) came from the Midwest. The tradition of the autodidact who later acquired formal learning is an old theme, especially calculated to warm the cockles of America's populist heart. But the plains did more.

Its northern naturalists recast their experiences into formal learning, just as artists and litterateurs did. The upshot was an American school of botany, which segued into a school of ecology. With uncanny fidelity, its life cycle meshes with that of the pioneering era, which is to say, with the epoch of quasi-wild prairie fire. As their chronicler observes, "at the core of their community was the shared experience of the prairies and plains." The first generation grew up on the frontier, studied botany as farmers commenced serious sod busting, and "created the science of ecology" at the same time that American philosophers, also responding to the vigor of nation building, fashioned pragmatism. The second generation—their students—could, as Ronald Tobey describes, still "experience isolated fragments of pristine prairie" and "feel the entire presence of the original plains as a deepening echo in their lives."[2]

Once again, a contribution cannot be separated from time and place. The "first coherent group of ecologists" in the United States emerged from the study of grasslands. The model of the natural world that they constructed exactly mirrored the experience of change that they had grown up with. Nature, too, was a pioneer, whose progress led to a civilized and stable climax, and while periods of rapid transition were disruptive, those disruptions only reset the clock. The Great Plains were too vast and implacable for humanity to deform in any serious way; at best, they could substitute domesticated grasses for wild, and livestock for bison and pronghorn. Most felt, as John Weaver, leader of the second generation, put it, that the prairie "approaches the eternal."[3]

The project began with Charles Bessey, who moved from Iowa State to the University of Nebraska in 1884, introduced laboratory methods into the classroom, wrote the dominant textbook on botany, and inspired legions of students. Bessey was a pragmatist (what we know is what we can learn from experiment) and a pioneer (we have to do in order to know). His two students, with whom he constituted a first generation, Roscoe Pound and Frederic Clements, split that inheritance. Pound helped bring pragmatism to jurisprudence. Clements introduced the quadrant to ecology, which was a practical means of breaking down the undifferentiated vastness of the plains into study units—the quadrant as the sodbuster plow of prairie science. Having devised methods, Clements ventured into other landscapes, notably, the Rockies. In 1910 he produced America's first formal paper on forest fire ecology, "Fire History of Lodgepole Burn Forests." In a tidy reversal, the grasslands did not derive from the forest—they were not a biota with the trees removed; they were the model for how forests worked, at least when subjected to fire. Fire behaved in lodgepole pine as it did in tallgrass prairie.

The difficulty came when Clements and his successors translated their data into a paradigm of succession in which ecology could explain phenomenon by placing them into various cycles and epicycles of progressive change. It was, in truth, the ecological equivalent of William Morris Davis's model for geomorphology, which also underwrote an American school. The concept spread like red brome. The foundational observation was, as Tobey puts it, "the central fact" about the prairie was "its natural stability and tough perseverance."[4]

Then came the 1930s. Drought crushed even native prairie, which presumably had long ago adapted to plains climate. Drought and depression, moreover, not only stopped pioneering, they undid it. But not before the record of untrammeled sod busting had ruined the capacity of the prairie to prevail. The Clementsian model could account for neither nature's nor humanity's effects. Intellectually, the theory went bankrupt; and just as damning, it could not say how to intervene and correct the errors that had contributed. Students interested in grasslands migrated into range science. The theory persisted well into the 1960s, but primarily because its old members, ensconced in universities and research institutes, continued to publish within its framework and because it was easy to teach. It resembled a massive tree that could be felled but not uprooted, and whose stump was left to rot away. As the adage puts it, science advances one funeral at a time.

When the American fire revolution burst onto the scene in the 1960s, its partisans were dismissive of Clementsian ecology, which could only envision fire as a disruption of natural progression, not as a necessary process of continual renewal; and they were skeptical of academic science, which had so long supported it in defiance of common experience. When fire science rebuilt, it arose out of government labs and looked to the conifer forests of the Northern Rockies, the chaparral and sequoia of California, and the longleaf and rough of Florida. Prairie fire ecology became a subset of range management. And like prescribed fire, which struggled to find enough extra grass to support itself, the science of grassland fire scrambled to find sufficient intellectual fodder and space to free-burn. Instead it found refugia like the Curtis Prairie and Konza Prairie and the burning brush around Texas Tech.

For a while grassland fire provided the model for an American science. For much of the past century, however, it has struggled to find an adequate place for itself within science at all.

IN POLITICS

When drought fused with depression, the consequences were felt beyond farms and schools of ecology. Critics of an American way that had ripped up the national economy as it did prairie sod had many examples of environmental havoc whose costs were too great for laissez-faire responses; but none rivaled the Dust Bowl. With subcontinental dust storms blowing to the national capital, the Great Plains returned to center stage of a national debate about conservation and the relative merits of public and private economies. The outcome stood the saga of pioneering on its head.

The Great Drought was not an era of massive fires on the plains because generous fuels require abundant rain. The big burns migrated west—Matilija (1932), Tillamook (1933), Selway (1934). Their smoke palls were the forestry equivalent to the enveloping soil squalls. Only indirectly, by mobilizing national attention to the wastage of natural resources and by refuting Clementsian ecology, did the plains contribute to America's fire narrative. *The Plow That Broke the Plains* had its counterpart in the unfilmed, but widely understood, axe that broke the woods. Smoke and dust were their dark-double offspring.

If anything, the Dust Bowl experience (as synecdoche for the 1930s Great Plains) moved the region out of the nation's pyrogeographic focus. In 1947 a quarter million acres burned in Maine and another quarter million in South Dakota. The Maine fires folded into other alarms, helped nudge the country towards a formal civil defense program, and led directly to the Northeastern States Forest Fire Protection Compact. The South Dakota fires blew away with the wind. The Maine outbreak looked to the future. The South Dakota conflagrations echoed the past, as though a macabre reenactment of pioneering days.[5]

Yet the Great Plains were not through with fire, and the national fire narrative had not fully abandoned the plains. The mechanisms for return were two, and they differed in their focus, their methods, and their means of propagation. One was interested in restoring prairie. It had powerful cultural capital to draw upon and a deep well of empirical evidence that fire was elemental to any attempt at restoration. Whatever prevailing theories intoned, it was instantly clear to those up to their elbows in big bluestem and black-eyed Susan that prairies could only thrive if burned. As more prairie was protected, the more fire had to return. The model of controlled burning to advance grasslands, however, ran into stiff headwinds. However much humanity's firepower—for all our existence as a species, really—had depended on grasslands, for however long hunters and herders knew that game and livestock gathered around burned patches, which exerted a gravitational pull akin to Jupiter, moderns had lost that connection or had immigrated from places for which substitutions had been devised or had so committed to maximal production that they left no grass sufficient to carry fire or simply feared its wild extravagance. The preserved prairies needed fire, and they needed legitimacy for the idea of fire. They propagated that message through a civil society of environmental groups, especially the Nature Conservancy (TNC).

The second mechanism was the academic refugia created by Henry Wright at Texas Tech. The Wright group also promoted fire to restore grass, but the target at the end of their driptorches was brush. Their leverage came not from appeals to prairie but to paychecks. Fire was a relatively cheap, effective, and environmentally benign tool to drive back the mesquite and general scrub that overran pasture. The Wright staff resembled a prairie dog village that, through offspring, colonized new lands, particularly in the southern plains. Henry Wright codified his knowledge in his coauthored text, *Fire Ecology*, but its message was more often spread through extension agents. It appealed to ranchers: it made fire a legitimate tool in the barn. Across fire's fulcrum it balanced the ecological enthusiasms of prairie restorationists with the economic pressures of ranching. Later, the concept of patch-burning helped bridge economic interests with ecological benefits.

Prairie fire thus entered into America's fire revolution. That reformation needed more exemplars than working landscapes in the Southeast; and prairie restorationists needed intellectual heft from an emerging science of fire ecology and practical instruction from experienced burners. The Curtis Prairie at the University of Wisconsin and the Konza Prairie under the direction of Kansas State University evolved from research plots to provide some of the fundamental science and to announce demonstration sites. Fire and prairie alloyed with special force, however, with the emergence of the Nature Conservancy as a presence. Thanks to Katharine Ordway, TNC acquired significant holdings, from which they were reborn after a period of hesitation into apostles for burning.

TNC only became a national player in fire after the organization moved into Florida, where prescribed burning was not simply an implement for restoration but a means of survival. From its Florida chrysalis a national fire program emerged, bulked up, hardened, committed to National Wildfire Coordinating Group (NWCG) standards. That was an investment most private landowners, and even most TNC prairie managers, thought unnecessary. But with marginal federal holdings on the plains, some other institution (or institutions) would have to overcome the tendency of life on the plains to disperse and attenuate. Something had to thicken what the unbroken vistas thinned. Flint Hill ranchers and wildlife refuges held fire in niches, but TNC became the institutional fulcrum by which to leverage prairie fire into national awareness. TNC accepted that role, set up fire learning networks and training exchanges, and stiffened prescribed fire associations.

Even as his idealized model was faltering, Frederic Clements had argued vehemently on behalf of New Deal reforms and government intervention. That didn't happen—couldn't happen. The Shelterbelt soon deflated, even as its counterpart, the Ponderosa Way in California (a 650-mile-long fuelbreak), cut across the Sierra Nevada. The federal government bought abandoned or degraded lands and created national grasslands, which it placed under the administration of the Forest Service. But such lands were not large enough to reshape the economy of the plains nor to tilt the national narrative. Prairie fire was a local, not a national, event. Fire management would require alliances among small and dispersed sites under all levels of government and mixed ownerships.

By the time prairie restoration had become popular, the country had moved ahead. It had rechartered its fire programs and rewritten the country's master fire narrative. Burning prairie was not a primary means to change national thinking on fire. National thinking had reformed from other causes and was now applied to prairies. So, too, its long and expansive association with burning prairie had granted the Great Plains a distinctive fire culture that in places had persisted and, at certain seasons in the almanac of American history, had contributed to a national culture of fire. No other region could make so broad a claim.

Yet, in the end, with so much land lost to fire from plow and hoof, and with so little symbolic or charismatic value beyond the region, the plains found themselves outside the triangle that defined America's pyrogeography. Florida, California, the Northern Rockies—what happened here had national ramifications. What happened in the plains remained in the plains. Like a whiff of smoke that triggers deep memories, the Little Burn on the Prairie could invoke a wave of national nostalgia, but little more.

PLAINS, GREAT AND SMALL

PLEISTOCENE MEETS PYROCENE

UNNUMBERED PRAIRIE POTHOLES dot the northern Great Plains. Most are pinpricks or blotches. Many are ponds, and some, lakes. They dapple the moraines and loess lands of North Dakota like a geologic tickertape recording for nature's economy the rise and fall of Pleistocene ice. Eastern South Dakota has perhaps a million, sprinkling the landscape like oregano. Minnesota's prairies have their 10,000 lakes as much as the northern woods. Since they first appeared in the wake of the Wisconsin ice sheet, potholes in the prairies, they have been an indispensable waystation in the great flyway that seasonally sends flocks of mallards, blue-wing teals, blue and Canadian geese, ibis, egrets, cranes, pelicans, gadwells, and others, north and south in a biotic echo of the advance and retreat of the ice. For the traveling flocks the potholes are not bumps in the road: they are the caravanserai that make travel possible.[1]

It might, at first honk, seem odd that amid speckled wetlands there should be a role for fire. But enmeshed in tall- and mixed-grass prairie, there is no way that they could not burn along their edges since the surrounding prairie will burn routinely, and that, when dry spells drop the shoreline, they would not flare and smolder into the sedges and organic soils piled there, and that, when deep drought empties the waters and drains the peaty bottoms, the fires would not from time to time gnaw into and even scour out those cauldrons of peaty combustibles. Yet just as the birds' habitat requires both upland and wetland, so fire and water become the yin and yang of the pothole province. Those innumerable water pockets make a pyric scatter diagram: through it one can trace the curve of an unusual multivariate relationship inscribed across the American fire scene. They are fire's flyway over the plains.

The upshot is a tableau of fire and plains. The way that fire interacts with climate and grazing to make prairie. The way conservation must mingle public purpose with private lands. The fusion of the ancient with the evanescent, as a biome primed for almost explosive change rests atop a relic geology fashioned by the movements of ice sheets across tens of millennia. The dispersed character of the potholes, so scattered that mere fragmentation pales beside it and cohesion only appears in spasms (even on the wing). Coherence demands relentless exertion. Amid this starry field of potholes spangling the prairie firmament, patterns emerge over wide expanses, among them the biotic constellations traced by the migration of herds and flocks. In the prairie potholes fire management builds on the outwash of the Pleistocene.[2]

Arrowwood National Wildlife Refuge (NWR), 30 miles north of Jamestown, North Dakota, is large, old, and characteristic.

Settlers moved in during the 1870s as the plains tribes moved out. It was a harsh, remote land, given to a mixed farm-ranch economy. More settlers filled in the blank spaces during the early 20th century, but much native grass remained. When the 1934 Taylor Grazing Act ended the transfer of public domain to private hands, drought and Depression were ravaging what biomes had survived. Through purchase, tax delinquency, or resettlement, the federal government began acquiring lands for the protection of the Great Plains flyway—a part of reclamation programs that throughout the plains included the better-known Shelterbelt. It was not enough to cease the ruinous practices of hilly sod busting, overgrazing, and overhunting that had damaged the prairie. It was necessary to rehabilitate them.

In 1935 Arrowwood Wildlife Refuge was established out of the former Riebe ranch, homesteaded in 1882. Together the Works Progress Administration and Civilian Conservation Corps (CCC) Camp 2774 put in dikes to help stabilize the wetlands, erected housing for staff and equipment, planted shrubs and trees (including chokeberry and Russian olive), and controlled predators. They built trails and roads that doubled for fuelbreaks; and when fires broke out, they fought them. The war returned management back into custodial status. The postwar era saw a revival of applied science, an agronomic model, to maximize waterfowl production.[3]

But the system was not quite right, and those close to the land realized that the number of ducks shot was not a measure of habitat health any more than board feet measured forests or animal-use months measured grasslands. They needed more land for coherent habitats, and they had to think in ecological terms. The land issue was addressed through purchase and a program of easements on private land for wetlands and grasslands. Owners could still farm and graze, but they couldn't break new sod or drain off the waters. The upshot was a density of mixed-ownership protected sites unlike anything elsewhere: North Dakota has 63 of the 561 units in the National Refuge System (the nearest rivals are California with 34, Florida with 30, and Louisiana with 23). But when combined with auxiliary easement wetland management districts and waterfowl production areas, they render the Missouri and prairie coteau into a veritable Milky Way of protected sites. Arrowwood was an administrative complex that included three fee-title refuges, six easement refuges, and three wetland management districts.

Still, more land would be ineffective unless it was managed properly, and not draining or over-shooting was not by itself sufficient management. The wetlands were only as good as their adjacent uplands, which were habitat for many species other than waterfowl (such as prairie chickens), and that biome was in bad shape. Something was still out of whack. In 1963 Arne Kruse and Leo Kirsch, later joined by Ken Higgins, began experiments at Woodworth Study Area (subsequently the Northern Prairie Wildlife Research Station) to see if fire could defibrillate a habitat not only decadent but becoming infested with exotics like smooth brome and Kentucky bluegrass and woody plants that intercepted ground water and broke the hydrology of wetlands. They used Arrowwood refuge for their beta field trials. They found that fire was inextricable—and indispensable—to good prairie and hence to the long-term health of waterfowl and upland birds. Their trials thus paralleled the fire revolution, and in 1972 Kirsch and Kruse addressed the 12th Tall Timbers Fire Ecology Conference with their findings. The gist was that fire's removal "has done untold damage to prairie wildlife." What was good for the prairie was good for birds.[4]

In this way plains waterfowl were inscribed into the honor roll of firebirds that had, over the course of the 20th century, forced a reconsideration of landscape fire. The collapse of red grouse in Scotland, as grouse replaced sheep and protected covert replaced widespread burning, had inspired several parliamentary inquiries. The decline of bobwhite quail in the coastal plains of the Southeast had sparked the Cooperative Quail Study in the 1920s. In more recent decades the endangered red-cockaded woodpecker has brought another fire-drenched habitat to public attention. Between the prewar quail and the postwar woodpecker came the birds of the plains—the migratory waterfowl and the prairie chickens primarily. They all—every study—paired sagging populations with deteriorating habitats, and they each, every one, recognized that the loss of birds followed a loss of fire. In its final 1911 report the Grouse Commission noted that the diseased grouse were also a harbinger of ill sheep, which is to say of sick heath, and it concluded that landowners needed to burn big, perhaps a third of an estate annually. Its successors agreed.[5]

This was, however, a hard sell to an agency historically committed to maximizing game birds because it meant a short term loss of ideal conditions, which for ducks meant cover and nesting conditions, for the longer term health of the habitat. The wildlife community thus aligned with ranchers who refused to lose any forage to fire, even seasonally, if it meant less stocking. But the value of the wetlands depended on the vigor of the upland prairies, and that meant serious burning.

The U.S. Fish and Wildlife Service (FWS) had plenty of fire experience, but like its lands, that knowledge was local, diffused, and specific to units. The agency had no national fire program, not even a coordinator. That began to end after fires in the Kenai NWR became explosively expensive; after fires in 1976 at Seney NWR in the Upper Peninsula—one natural, one set—went wild; and then yielded to a series of fires that were either botched or resulted in deaths. The climax came in 1981 when two technicians died fighting a fire at Merritt Island NWR. Congress raked the agency over its mismanaged coals. But out of that ordeal by fire, the FWS got marching orders and money, and despite the agency's inertia it began to integrate with the national fire community, which meant adopting national standards.

One unexpected outcome was that, in the prairie pothole region, the FWS became the largest federal fire presence. The U.S. Forest Service (USFS) had a few national grasslands and remained the conduit to the states, but in places like North Dakota the FWS could do with fire what it had done through easements and collaboration with wildlife protection. What migrating flocks did to bind the pieces together ecologically, the FWS had to attempt institutionally.

Bonding to national trends brought a more professional fire staff, but it also welded refuges to national cycles of fire management. These boomed after 1994 and boosted again after the 2000 National Fire Plan. Then they collapsed as Congress and the Great Recession applied fiscal tourniquets to the public sector. In 2011 the region had a permanent fire staff of 58; by 2014, 26. Funding collapsed, which meant projects needed partners, and with most grasslands in private hands, that meant public-private collaboration at a time when rural populations were collapsing and there was little economic incentive to burn. Few ranchers would sacrifice immediate forage for sustained grassland health, even as cool-season invasives like Kentucky bluegrass and smooth brome shrank pasture. There was no Conservation Reserve Program for fuels. There was no endangered process act for fire.

Even so, and after rising and falling from the 1970s, burning became a routine practice: it was reestablished along with big bluestem and sharp-tailed grouse. The Dakotas prescribe burn 40,000 acres a year; Arrowwood alone burns an average of 8,000 acres annually, or roughly 10 percent of the total land under its administration. Yet this is no more than a tithe on what the prairie needs, and the acres burned are a weak index of the mix of burns, dormant and growing season, back-to-back and spread out, surface flushing and deep scouring that the land needs, and all this must be syncopated with grazing to achieve the best results. Prescribed burning and prescribed grazing are the prairie's winning tag team.

After the flush from early burns in the 1970s, the ecological effects have worn out; nothing persists long unchanged in the prairie. Fires that had come no more than 4 years apart must now wait 44. Amid declining budgets and staff, and worse, political instability; with a warming climate shuffling northward and exotic weeds seizing the better, formerly disturbed soils and massed at the borders, ready for invasion; with surrounding buffer grasslands put to the plow to grow corn and soya, a new era of sod busting accelerated as crop insurance encouraging putting Conservation Reserve Program and marginal lands into production; with critics finding more imaginative causes with which to restrict the scope of burning, whether it be smoke or an eccentric butterfly, refuge managers are struggling to keep enough good fire on site, even as it becomes ever more apparent that fire is the strong nuclear force that bonds the biotic pieces into something like ecological integrity on a land that needs as much resilience as it can get if it is to absorb the still greater blows to come.

The problems can seem as dense and diverse as the botanical mix of grasses and forbs, microbes and mammals, insects and birds, that constitutes prairie. But boil that pitchy brew to its essence and it leaves a hard distillate called scale. The glory of the prairie pothole landscape—its expansive range, its role as dispersed way stations of a flyway that spans North America—is also its curse.

It includes the problem of small units that need burning—a question of edge to area. It takes as much effort to burn 200 acres as 2,000. The diffuse character of private and public land ownership means it isn't possible to subsume hundreds of upland-wetland units into a single burn, or to muster the varied entities and folk into common set-piece undertakings. Without whole landscape burning, fire officers can't muster favorable economies of scale. They can't mass critical capacity and infrastructure.

Scale means that they can't trade space for time, or time for space. Few sites have sufficient bulk to both patch burn and let bison or cattle free-range. They can afford to burn land that temporarily reduces forage for livestock and nesting sites for birds only if they have enough land elsewhere to compensate. They see fire as a tool, not as a necessary process that must be fed and tended. They can't accommodate the quirky needs of the Dakota skipper butterfly, which thrives in older prairie, with the needs of plovers, which flourish in recent burns. Instead of expanding and absorbing diversity, each new demand shrinks the room for maneuvering, so that the cost of operations becomes exponential, and the price of failure irreversible. The grasslands move quickly: without room to roam, it's hard to buffer and balance conflicting needs. Arrowwood NWR has perhaps 16 days a year available for burning. Today, the rate of change, always sudden, is becoming explosive.

Scale means it isn't possible to maximize management for one value or another. The parts will only flourish within the context of the whole, but the whole demands more space to accommodate its complexity. To acquire the resilience needed to survive, you have to manage for the habitat overall, and that requires enough land to accommodate the peculiar needs of all the parts and processes. The Great Plains as a biota flourished because there was space enough to move and seize the particular opportunities of varied sites and seasons. Always, some parts were out of sync, yet the whole was hale. When those once-expansive habitat smorgasbords shrank to the size of teacups and teaspoons, the capacity to absorb and respond became brittle and vanishingly weak. There is less margin for error, less capacity to counter threats, fewer opportunities to reclaim lost ground.

Nor is it enough to put fire as fire on the ground. Vigorous prairie needs lots of fires in many combinations, each with its own compounding interactions with grasshoppers and bison. Arrowwood needs a medley of fires to match its menagerie of waterfowl. It needs double burns in a year to stall Russian olive, repeated spring burns to hinder Kentucky bluegrass, fast burns in the fall to hammer snowberry, and burns year after year to promote bluestem and grama, with places and times spared flame for a spell. Yet it is far worse to have too little fire than too much. Almost any regimen for fire on prairie is better than no fire. And a fire regime once extinguished is, like an extinct predator, tricky to return. The same is true for a fire culture.

There is plenty of room for despair or just plain cynicism. But there is always hope. In *Prometheus Bound*, Aeschylus has his tortured but still defiant Titan declare that, along with fire, he gave humanity hope. Those who keep the flame are keeping the future.

Today the northern plains are increasingly split between two realms of combustion that symbolize two geologic epochs. One is a relic of the past, and its flame is kept to preserve the living memory of that past. The other is a harbinger of the future, and its flame threatens to remake the planet. The first powers migratory North American wildlife; the second, an ever-restless American society.

The prairie pothole region is a relic world, a geomorphic imprint from the Ice Age. For millennia flames swept the region, sparing only the wet blotches and potholes themselves, the speckled traces of the old ice. Today's residual burning is diminished, like dispersed candles compared to the sweeping conflagrations of the past. Yet the biota, and ultimately the potholes themselves, cannot survive without those surface burns. The flames migrate with the seasons like geese and bison.

The coming world is the glowing constellation of gas flares that map the Bakken shale being worked for oil, an emissary from a gathering Fire Age called the Anthropocene. Drill holes belching natural gas speckle western North Dakota in an eerie counterpoint to the pothole country to the east—a dark pyric double. These fires burn day and night, summer and winter, without regard to surroundings; they propagate along a flaming front of fracking. They announce a future world shaped by fire as the past was by ice.

Those flares are not static. In reality they are combusting downward into the rising gases, and in so doing they are burning down into the geologic past, while their emissions loft upward into the future. That industrial firepower is rewiring the American landscape, redefining what of it is a resource and recoding how those resources might be routed through a human economy. They are shifting climate, allowing for the northward spread of crop cultivation. They power the tractors that plow under the prairie. They run the distilleries that process corn and soy into commodities and then haul those products to market. They run the motor homes that send Dakota snowbirds southward for winter. They make possible a new biochemical era of sod busting that relies on Roundup and Roundup-ready seeds to replace native flora with genetically engineered cultivars. Just as the upland burns at Arrowwood are a catalyst for a whole biota, so the fracking flares are for America's industrial society.

The prairie potholes are caught between those two fires. They desperately need more of the first and far less of the second. But how the prairie potholes and the prairie frackholes play out will determine what kind of future we might hope for. In this tension, too, they are a portal for realms of fire for the Great Plains overall.

ECOTONE

WIND CAVE NATIONAL PARK is an instant anomaly: a celebrated firescape on land preserved for its subterranean caverns. What makes its fire scene compelling is the almost perfect balance between mixed-grass prairie and ponderosa pine forest. What makes its fire story significant is how that ancient quarrel played out during America's fire revolution. Its biogeography makes Wind Cave a classic ecotone. Its evolving fire practices made it a historical one as well.[1]

The landscape works, ecologically, to keep that balance between prairie and ponderosa. The park lies on the southeast, lee side of the Black Hills, windy but not too windy; storms break on the north and west. The landscape burns frequently, but not too frequently; perhaps 10 to 12 years on average, with a range from 2 to 34. Its fires are large, but not too large; 600 to 6,000 acres within the boundaries. At 28,500 acres the park is larger than its fires.[2]

The fires come routinely enough to thin out pine thickets and kill pockets of mature trees but not hard and often enough to abrade the forest away. In the lowlands to the east the scene is grassy, a fire steppe grading into a pine savanna; along the western summits, it thickens into a forest with a serviceable understory of grass. The pine tumble down the ridges like a biotic rockfall into woody talus and stringers and scattered clumps. A serrated, even fractal fringe etches the frontier between mix-grass prairie and ponderosa forest. Park fire staff estimate the relative ratio of grasses to trees as 50:50.

What further checks the fires is a full complement of indigenous fauna that compete for, break up, and sculpt the fire-bearing grasses. The park was created in 1903 to celebrate an underground curiosity. But its surface landscape, an intact prairie, was ideal for restoring grassland fauna. In 1912 a game preserve, under the direction of the U.S. Biological Survey, was established adjacent to it and stocked with bison, elk, and pronghorn. The preserve was subsequently transferred to park jurisdiction. The land already held most prairie mammals, including extensive prairie dog towns. For predators it had cougars, wildcats, and (later, introduced) black-footed ferrets. All it lacked were grizzlies and wolves.

The outcome is one of the purest expressions of a prairie-pine ecotone anywhere. The Black Hills are a western firescape, but plucking them onto the plains makes their flanks into a hybrid in which grassland and forest spill into each other. The resulting biota is probably as close to presettlement conditions as anything in the Great Plains. The landscape remains an ecotone.

The last landscape-scale fire in the region occurred in 1881 before the usual suspects—logging, grazing, fire suppression—crashed the system. That was the western side (the woody, so to speak) of the park's historical ecotone.

Its grassy, eastern side, as it were, spared the park itself from the worst excesses. It was a small polygon in the regional matrix, was never logged over, shed its livestock grazing between 1906 and the early 1930s, never had a serious capacity to fight fire until the CCC arrived in 1933, and then, with lightning kindling many starts, its rich grasslands meant fires could burst through for a burning period or two before the winds paused or crews could muster. Its largest recorded lightning fire occurred in 1961 (1,156 acres). A year after the Leopold Report, a cigarette-kindled blaze burned 5,468 acres in the park, and 13,000 overall.

In brief, the land never lost its fires, or at least kept enough to prevent the biota from spiraling toward one extreme or another. When the fire revolution arrived, the park had anchor points from which to reintroduce prescribed burning. It also had a cornucopia of advantages: nearly every feature of its circumstances favored keeping fire. It still had ample grass, with few invasives. It had a charismatic creature, the bison, that fed on grass. It held no significant cultural artifacts that fire might impact. It had no wildland-urban interface. It housed no threatened or endangered species. It faced no concerns over smoke. It was mostly bordered by public lands; its private-land neighbors lay within the Black Hills Fire Protection District; and in 1973 it joined the Interagency Prescribed Burning Coordination Committee, a move that helped overcome isolation and problems of institutional scale. It boasted a landscape with a high degree of ecological integrity and resilience, one able to absorb wildfires into its fire management regimen. Its signature feature, Wind Cave, lay well below fire and smoke. It had access to research capabilities from the Forest Service and South Dakota State University. Its governing bureau, the National Park Service, was keen to reinstate fire, which it saw as a natural process, and was tolerant of fire's potential interaction with other disturbances, such as mountain pine beetle.

Wind Cave had political and geographic space in which to act. It was big enough to do something, yet not big enough to attract unwanted attention. It had none of the killer-app issues that flattened many programs. It was not a celebrity landscape guaranteed to draw quirky critics. It straddled a historical ecotone in which its fire program might side with the revolution or elect to dawdle. It lay with Wind Cave to choose.

It chose grass and fire. In 1973 it commissioned research under an agreement with the U.S. Forest Service and South Dakota State University, and commenced experimental burns, which seamlessly segued into an operational program that balanced the ecotone between woods and prairie. It moved choice into action, and then it has continued the program for 40 years. That experience (and relative heft) has made Wind Cave the center of fire operations for seven National Park Service sites throughout the Black Hills and Badlands.

However auspicious the circumstances, that transition did not happen on its own. Conditions were favorable only if someone acted on them; left to themselves, the environmental pressures favored the woods, and bureaucratic pressures favored the choice of making no choice. On one side of history's ecotone stretched the relentlessly encroaching past of fire exclusion, and on the other, a still-open future of fire restoration. Many sites, big and small, both the known and the obscure, found themselves at that similar edge and chose poorly or, more often, simply found ways to dally and fret until the forest had pushed the fringe of history back into the past. A few sites picked up the torch, opened up new grassland, but then faltered in trying to keep the torch lit, and the woods returned.

Wind Cave chose the grass, it chose early, and it has kept choosing. The rest, as they say, is history—but history only if we appreciate how choosing the future becomes the past.[3]

NIOBRARA

THE NIOBRARA RIVER WENDS the north flank of the Sandhills. Trees flourish in the floodplain and along the deep, if gentle, embankments. But to call the Niobrara Valley simply a gallery forest amid the plains is to dismiss the Smithsonian as just a museum. Perhaps more than any other locale, it claims a biotic intersection of north, south, east, and west.[1]

The river crosses the traditional divide between humid east and arid west. It marks the western limit of 83 eastern species and the eastern limit of 47 western. It contains a sampling of the Rocky Mountain biota, including ponderosa pine and juniper; but the bottomlands grow American elm, bur oak, ash, and other eastern deciduous species. All of these species belong with the biota of southern North America. On the north-facing slopes of the valley, the biota of northern North America—elements of the boreal forest—take root. All three variants of Great Plains prairie grasses—short, mixed, and tall—thrive in local niches. The Niobrara has both aspen and birch, club mosses and serviceberry, little bluestem and grama.

All this is well known and is why the valley has been protected. Less well appreciated is how the Niobrara marks an institutional crossways almost as dense. Free-range bison mingle with herded cattle; private land with public; preservation with profit. The interaction of natural conditions and human land use places the Niobrara squarely at the center of the Great Plains matrix for fire. The biota argues for burning. The history of settlement testifies to fire exclusion. Contemporary management wends between the cliffs of national and local interests, the fear of fire wild and the promise of fire prescribed, the power of fire to catalyze a more robust landscape and the need to restore a fragile one.

The bison and Plains Indians left by the 1870s. Cattle and ranchers moved in. For fire the chief facts were suppression and the substitution of domestic cattle for the indigenous grazers. As the herding economy sagged, old ranchers died out, and proposals for a dam arose, the Nature Conservancy intervened to commence the modern era in which the primary values of the valley ceased to center on commodities and the narrative no longer continued the saga of pioneering.

In response many levels of ownership and governance with an interest in conservation converged on the Niobrara. In 1980 TNC purchased two major ranches, and in 1985 began replacing cattle with bison. A decade later it added more ranches to make the 58,000-acre Niobrara Valley Preserve. The State of Nebraska acquired a park (at Smith Falls) and three wildlife management areas. The U.S. Fish and Wildlife Service anchored the upstream valley, where the Minnechaduza Creek joins the Niobrara with its Fort Niobrara National Wildlife Refuge. The main valley remained at risk, however, until 1991, when Congress included 76 miles of the Niobrara within the National Wild and Scenic River System under the administration of the National Park Service. To expand the range of protection, the Nature Conservancy negotiated conservation easements with several ranchers on the north slope; the preserve proper helped pay for itself by leasing many uplands south of the valley for cattle. The rest of the landscape remained in private hands, some for ranching, some for recreational use.

Each of these landholdings was responsible for fire management, and one might expect that fire management might also organize collectively. The only force of any power, however, were the volunteer fire departments (VFDs) of the surrounding communities, notably Valentine, Johnstown, and Ainsworth. They were, of course, volunteers, their mission was suppression, and they had little interest or time to engage with a full spectrum of fire-management projects. One might expect that the federal agencies could call up sufficient resources, and they could, although not in time or in quantities adequate to their perceived needs. TNC had considerable fire expertise but only a tiny staff at the preserve, insufficient to conduct operations of any complexity. Both the feds and TNC, moreover, had to meet NWCG standards to conduct either firefighting or fire lighting. The local ranchers who traditionally assisted (two of whom were hired to manage the cattle and bison herds) did not qualify, and neither did the VFDs, which reduced capacity even further. In 2006 the Niobrara Valley Prescribed Fire Association was organized, which included fully furnished burn trailers from Pheasants Forever, but since members were not trained to NWCG standards, neither the feds nor TNC could use them.

Fire management could only succeed if the various agencies pooled resources. But there are fewer staff than meet the eye, and collecting them under a common cause, particularly one not fully accepted by all members, makes prescribed fire difficult. Even TNC, which needs to burn (and is avid to do so), has to request permits from the local VFD chief, and if the burn lies within a quarter mile of the Niobrara River, from the National Park Service as well. This restricts the slickest solution to controlling a burn: to drive the flames into the river. The only easily accepted fire, perversely, is wildfire.

There is little doubt that fire was frequent in presettlement times—how could it be otherwise?—that it burned across seasons, and that its range has shrunk over the past century.

The prevalence of fires was a common event in settler chronicles. The seasonal mix of burning prevented one type of fire from dominating. And the loss of that scale and diversity is what best characterizes the shift in fire regime. Studies of fire-scarred pine nearby yield an average return interval from 1857 to 1900 of four years, from 1901 to 1950 of five years, and from 1951 to 1985 of six years. That's a pretty rapid historic return and only a slight shift compared to the collapse typical of the plains overall. Throughout, fires continued from lightning and accident, and in 2006 a broken power line sent fire roaring into Valentine with the loss of 12 houses.[2]

Yet the ecological effects of fire's removal are clearly apparent. Cool-season grasses have invaded; forbs have diminished; and trees—especially the weedy eastern red cedar—have spread widely. Photos of the Niobrara

Valley taken a century ago show patchy woods in what today are continuous and dense mixed forests. Trees have spilled over the valley rim like a river overflowing its levees. The exact dynamics are tricky since the reckoning must factor in the altered grazing regimes, for cattle and bison interact with burned prairie differently.

The preserve introduced bison and burning in 1985. The bison have grown to 600 heads split into an east and west herd. The burns were initially timid and tiny, many at night to avoid escapes and not alarm neighbors. While the bison hit the fresh grass hard, they moved, and many of the forbs remained to help hold the soil. Because of the complexity of the scene, the window for burning is small, and local VFD chiefs are quick to declare burn bans when conditions favor hot fires; but these are often what the preserve needs to sweep away its woody weeds. Cattle complicate the scene in another way because the summer grazing is conducted under lease (about 1,800 cow-calf pairs), and the preserve must guarantee adequate forage or find alternative pastures while keeping the cattle off the fresh burns. Ideally, fire management would like to burn 6,000 to 8,000 acres a year on a four- to five-year rotation. In practice, it burns an average of 1,000 to 1,100 acres.

Nothing unusual in those numbers—no one burns as much as they want, and the preserve does better than most and has made burning a culture. The reasons behind the slippage are the common ones; put simply, the plains disperse what management needs concentrated. While the preserve holds many pieces it also stretches over 26 miles, which means a lot of perimeter to hold. The best compromise at present is to conduct two-week burning programs in the spring through the Fire Learning Network for fire training and qualification renewal and then burn every day conditions permit. The cadres come from throughout the region and across agencies. In a curious way it's the institutional equivalent of migrant workers who in old days harvested wheat and rounded up cattle, and in a weird way, of patch burning and grazing, with fire staff taking the place of bison.

Environmentalists rightly celebrate biodiversity and note the need for managing on a landscape scale. Strategists of contemporary fire management likewise argue that scale is critical. Fire is too complicated and costly as boutique burns, and it needs space to create a full range of outcomes and buffer against escapes. The transactional costs of a small burn are often the same as for a large one. The only way to establish that scale of landscape and muster the capacity to operate is through cooperation with neighbors or a national pool. In principle, institutional diversity could complement biodiversity.

It doesn't often work that way. Complex landscapes require complicated programs; there is no single practice or restored process that will satisfy all the parts; each need is balanced against the others. So, too, diverse communities, while they can collectively amass the mind and muscle to do fire, require immense investments of attention to meet the particular desires of shareholders. Far easier is the circumstance with a single landowner, or public lands even when administered by several agencies. Fire-management communities don't self-organize into existence. But without them the only fire to do the ecological work is wildfire, which is a risky card to play in a high-stakes game of ecological integrity and the biodiversity of small preserves.

The burning goes on. The Niobrara Valley remains equally at a crossroads of continental ecology and the complex matrix by which a modern society must exercise stewardship over it. If it holds bits of a biota from all the cardinal points, so it also concentrates management concerns from all points. From the west (symbolically): restrictions on burning, from local fears to national standards. From the east: the need to muster capacity for prescribed fire, to tweak fire organizations founded on suppression to adapt to prescription burning. From the north: problems of seasonal timing and scales of burning. From the south: the integration of fire with fauna.

What makes the Niobrara Valley special is its capacity to amalgamate species whose biotic hearth lies elsewhere. Here they have gathered, as if by a process of ecological stream capture from distinct watersheds. But that varied ecology is matched by an administrative landscape of equal diversity, each piece of which has its thematic core elsewhere. Whether, at the Niobrara, those pieces can come together as a functioning whole or instead default into a community of convenience will determine what kind of fire the valley will experience.

SANDHILLS

And in their ears ran Jules's last command: "Watch your fires or you'll burn out the country."

—MARI SANDOZ, *OLD JULES* (1935)

THE SANDHILLS OF NEBRASKA constitute the largest contiguous grassland in America, the major recharge watershed for the Ogallala reservoir, a vista of unbroken sky and wind, and a landscape of fires whose return was once as common as migrating geese. Life in the Sandhills, as throughout the plains, circled around its grasses, and, as often as not, the grass cycled on fire—thus concludes the opening paragraph of Mari Sandoz's classic account of pioneering, *Old Jules*: "Fringes of yellow-green crept down the south slopes or ran brilliant emerald over the long, blackened strips left by the late prairie fires that burned unchallenged until the wind drove the flames upon their own ashes, or the snow fell."[1]

The pioneering narrative is the epic of the northern plains, a continuation of the colonization of the North Woods but with grass and hills in place of forests and lakes. Around the Great Lakes the task was to break down the woods and to plant domesticated grasses. Around the Great Plains it was to plow up the grass and, as a symbol of civilization, to plant trees. The trees were an essential task of reclamation, and many bore fruit, and all serve to shield against the wind. But whether or not they defied the wind, which they could not prevent, they defied the transience that seemed endemic to the American steppes. They stood for cultivation, for a (literally) rooted society, for a determination to resist and endure. Arbor Day began in Nebraska in 1872.

No one could build a stable society on sand, so if the Sandhills were to be cultivated, they would need trees. America's great experiment in afforestation began there in the early 20th century.

What became the Nebraska Sand Hills Tree Planting Project had many contributing causes and one catalytic personality. The upshot was America's first tree nursery under federal sponsorship and its first large-scale attempt to remake landscapes by planting forests.

The belief that trees were necessary was based on a global understanding of what forests did and how they assisted settlement. Want of "wood and water" was the great hardship of rural life in the prairie. Wood meant construction material, fuel, and amenities like shade and figuration on the landscape. Water was even more vital, and as settlement approached the zone of unreliable rainfall farmers turned to irrigation. Where water was ample, so were woods, and, it was argued, by the logic of sympathetic magic, if woods became abundant, so would water. Since the 17th century many traveling naturalists had commented on the obverse: that where forests had been cleared, the watershed destabilized and climate swung wildly between drought and deluge. The science of the day agreed; the observations were even formalized into what has been labeled a "desiccation discourse." Where forests could not be preserved, they had to be rebuilt. It seemed then that widespread afforestation could ameliorate the climate, or at least provide those material necessities settlers everywhere valued and those on prairies required.[2]

The earliest efforts dated back to the later 18th century, but quickened as Europe's naturalists followed its expanding imperium. The doctrine of "forest influences" (mostly their effect on climate) underpinned the campaign to create forest reservations in colonies. But it could easily be turned inside out to argue that one could overcome natural aridity by creating woods. On a less abstract level, forests were seen as practical means to encourage agriculture where the soils or climate were too harsh, or where bad practices such as overcutting, overgrazing, and promiscuous burning had stripped the forest's beneficence away. The scale—geographic and temporal—of such projects required state sponsorship. Prussia planted the sandy wastes of the Baltic to pine. France reclaimed the sand dunes of the Landes, and replanted across the Alps to prevent a succession of disastrous "torrents" that raged far downstream. America's first foresters were well acquainted with this literature, and so were others attuned to the early movements of what became a doctrine of conservation. Foresting the Sandhills was an American echo of Europe's exemplars. But someone had to promote the idea, and for the Sandhills that pivotal personality was Charles Bessey.

Bessey was no crank, no autodidact from the frontier fired with a prophetic quest, no promoter gilding commercial schemes with a patina of science. On the contrary, he had attended college in Michigan, written a founding textbook, *Botany for High Schools and Colleges* (1880), become a professor at Iowa Agricultural College, and then, after a long courtship, accepted the chairmanship of the Botany Department at the newly established University of Nebraska (1885), later advancing to dean of the Scientific College. Quickly he established a Botanical Survey of Nebraska (1892), founded the first lab-based course in botany, revised and leveraged his influential textbook into an American school, and inspired America's first suite of ecologists through his most famous student, Frederic Clements. The study of American grasslands began with Bessey, his famous botanical seminar, and his thousand-and-one inspired students who followed his exhortation to "missionize" for botany.[3]

Occupying a quarter of the state, the Sandhills loomed large in the Botanical Survey, naturally concerned those keen to promote the Nebraskan economy, which meant agriculture, and interested those like Bessey who were sensitive to the Progressive politics of the day and the urgency of so-called timber famines. At the Sandhills two observations most struck him. One was the discovery of ponderosa pine in tiny pockets, which he believed indicated that the Sandhills had previously been forested, certainly during the Pleistocene. The other was the spontaneous reclamation of prairie by trees. In fact, trees were advancing from both west (ponderosa pine) and east (deciduous hardwoods and cedar). Clearly, trees had once clothed the Sandhills and were busy reestablishing a presence. Why had they disappeared, and why were they returning? The best explanation was fire. Burn annually and trees would fail to survive. Remove fire, and not trample or cut what regenerated, and forests would reclaim their lost inheritance. One could accelerate the process by deliberate planting and by excluding fire.[4]

Bessey campaigned for a site to demonstrate the possibilities, and in 1902 President Theodore Roosevelt proclaimed two still-public patches of the Sandhills as forest reserves; in 1906 he added a third. The boundary surveys for the reserves quickly segued into the identification of a suitable site for a nursery along the Dismal River near Halsey. Crews collected ponderosa seeds from the Black Hills, and in 1903 began plowing, planting, and protecting. The initial results were, as Charles Scott reported, "very discouraging"—fewer than 10 percent survived the first summer—but the crews "still had faith in the project" and persevered.[5]

They redesigned the nursery, experimented widely with species, and planted. The 1912 Kincaid Act expanded the homestead acts and provided free seedlings to private landowners until it was succeeded by the Clarke-McNary Act, whose Conservation Trees Program dispersed seedlings to private landholders for large plantings and shelterbelts. In 1929 the first pruning of the plantation occurred. The advent of the Civilian Conservation Corps strengthened the infrastructure of the nursery and forest, by now named in honor of Bessey. The plantings and windbreaks pioneered along the Dismal River became the model for the immensely expanded Great Plains Shelterbelt project inaugurated in 1933. In 30 years the Bessey forest had gone from a project denounced by ranchers as "a crazy fool idea" to the proof-of-concept prototype for a national campaign to reclaim a region blasted into a dust bowl.

The schemes failed. Afforestation as a practice did not spread, although in the absence of chronic burning, trees continued to self-propagate; and while the Great Plains Shelterbelt soon faltered, smaller windbreaks took and, with disheartening irony, often became points of infection for feral trees that threatened to smother the range. But the Bessey forest survived. The Bessey Division of the Nebraska National Forest today boasts 30,000 acres of woods and 60,000 of prairie.

The experiment proved Charles Bessey right. Trees could grow if fires were kept out. Whatever the arboricultural challenges and however much ranchers, already loath to yield free grass to any other purpose, scorned, the operational fact remained that without fire protection the forest would be swept away.

The fires that had once crossed over the Sandhills like its bison and pronghorn continued. The Sioux left; and settlers, mostly from Europe, replaced them as fire-starters, but the old choreography that had brought fire, wildlife, grass, wind, and people lost its rhythms. There was more fuel and less controlled burning. Describing the Sandhills country west of the Bessey, Mari Sandoz recorded fires from "greenhorns" who carelessly stopped to smoke and set dry prairie aflame; from malicious cowboys and homesteaders attempting to burn out competitors; from locomotive smokestacks and brake shoes once the railroad arrived; from the sheer abrasion of newcomers on an old and dry-grassed land continually bending and sifting in the wind.

The settlers, she remarked, soon "knew about fires." In the Sandhills they responded, as they did across the plains, by plowing "fire guards," parallel strips "approximately eight or ten feet wide and sixteen feet apart" around land they wanted spared—their crops, their houses and gardens, their winter pasture. But of course the median strip would regrow, and unless regularly cleaned out, the guards would fail. Still, when fires broke out, Old Jules instructed the kids to lie down in the plowed strips and wait out the burn.

Smoke on the ridge was a call for settlers to rally—all settlers, again as throughout the plains. They galloped to the column on horseback; the Springlake hay crew "rattled off" in heavy wagons. There a gang plow with six horses cut line, while behind them "backfirers" set flames against the upturned dirt and "singed men" swung sacks and "old chaps' legs" to dampen the burnout. Two groups, separately, "closed in from the sides." They had "tapered the fire and finally headed it, after contesting every step of the sixty miles between the Burlington tracks" and the present head. The countryside swelled with sun-reddened smoke and the odor of burning grass. The fight against a prairie fire became a set piece of pioneering, even a communal ritual.[6]

What threatened sod farms and ranches also threatened plantations of pine. In the early years of Bessey, Charles Scott recalled a "dry thunder shower" that kindled three fires only a few miles distant, none of which spread into the plantings. In 1907 a fire from a locomotive threatened but was stopped by a vigorous firefight along plowed "fire guards" that encircled the forest. In March 1910 in an early expression of the great drought that underwrote fires throughout the summer across the northern tier of the United States, a fire that started 65 miles west of the Bessey burned into it for several hundred acres. But a standard of rigorous silviculture, selective grazing, and formal fire control mostly kept the menace away. In the common-wisdom logic of settlement, wild fires would go the way of wild bison and wild indigenes as cultivation intensified.[7]

On May 5, 1965, however, a lightning strike from a shallow cold front kindled a fire in grass west of the forest around 11:25 a.m. Conditions were ideal for rapid spread: warm, dry, gusty, unstable. Two crews watched the ignition; one arrived within five minutes, the other a few minutes later, but together they were unable to halt the flames. By nightfall the fire had blitzed over 20,000 acres, about half within the plantation. Before it ended a day later it had scorched 60,000 acres of grass and 30,000 of timber. As a postfire review noted, these climatic conditions are common, lacking only a suitable spark. On average April has only two days with lightning, and May, six. The old suite of causes from settlement times had faded.[8]

So fires came episodically. In 1972 it was the Mullen fire; in 1986 the Powderhorn; in 1995 the Good Friday fire; in 1999 the Thedford, which burned 78,000 acres and into the town of that name and killed one volunteer firefighter. In 2003 the McClaren fire threatened Valentine and the Niobrara reserves. In 2011 every red flag day had a fire, all started by people and all of which miraculously spared the forest. They started from fixing fences, from a harvest combine sparking beans, from cigarettes tossed to the roadside, from four-wheelers and catalytic converters left on dry grass by grouse hunters. The fires blistered across the countryside for a day, then, as the winds subsided, the flames died down—very much the pattern recorded in settlement times. The big fires, however, began from lightning, for which the tall trees were an open invitation.

And a giant pile of slash. Times had changed. The trees that settlement craved, a later age little cared for beyond aesthetics and historic curiosity. Expecting many would die, the seedlings had been planted too tightly and then grew too closely, which made for deep litter and extensive prunings, all of which lacked any market or roads if a market could be found. The cuttings were piled and burned and fire kept out by traditional fire suppression. When, someday, a full-throated fire enters, it might well sweep the scene back to its presettlement state.

After a fire, to prevent damage to the grasses, cattle are banned for a year. Some selective grazing occurs on fuelbreaks and around and through the forest, but for the most part the old alliance between fire, grass, and grazers has broken for the same reason that it shattered to the east. The Bessey forest was a cultivated woods with no more desire for free-burning fire than an apple orchard. It was a tree farm.

The Bessey forest is as unique a cultural feature as the Sandhills are a physiographic one. Both are historic relics carried into contemporary times. The Sandhills are a geomorphic fossil from the age of Pleistocene glaciations. The Bessey is a living fossil of a time when conservation meant supporting agriculture and when the tree on the plains was an emblem of land stewardship.

Today, environmentalism is more likely to convert cornfields into native prairie than plant prairie to pine. Both exercises are built landscapes—as much cultural monuments as Mount Rushmore to the west—that testify to the prevailing values of their times. Right-thinking environmentalists today as often as not see the spread of trees over prairie as an ecological infection in need of cauterization. They prefer the indigenous to the imported, and where they worry over agricultural landscapes, they see monoculture invasions like that of eastern red cedar as ruinous to biodiversity. Instead of planting trees, they would cut them.

Or perhaps more aptly burn them. Curiously, the survival of Bessey and that of tallgrass prairies like Nachusa will depend on how they coevolve with fire. The Bessey plantation needed—still needs—to keep fire out. The tallgrass prairies, the prairie potholes, the prairie farmlands on which game birds thrive and the prairie ranches strangled by brush and woody weeds all need to keep fire in. Such paradoxes remind us that fire regimes, too, are cultural constructions, an always-awkward negotiation between what a society imagines its ideal nature to be and what material nature will allow.

THE NEBRASKA

THE FIRES THAT the Great Plains encourage are not the fires modern settlement is designed to manage. The plains tend to disperse populations rather than collect them; they invite mobility, often as seasonal migration; they diffuse properties over space and time. Ancient fire regimes mimicked those characteristics. On the plains wildfires spread and wildfires stop with equal suddenness. Controlled fires occur within narrow windows of fuel and weather. These are not attributes for a society dedicated to bounded lands and fixed ownership, that seeks to maximize grass for production, that founded natural resource management on forests, that has distributed itself in small clusters and the occasional large one but not, west of the 100th meridian, with more than a scattered and tenuous claim.[1]

Of course there are examples of places with extensive, lightly inhabited lands that accommodate fire with relative ease. Alaska, Nevada, the Northern Rockies—all boast few people and plenty of fire, and the two features seemingly reconcile. In these instances, however, ownership resides overwhelmingly with the public, and policy points to a tolerance for (or encouragement of) large fires. By contrast, the plains—excepting the West River country of South Dakota—lack that scale of public lands and governmental presence, and fires must burn on working landscapes, however sparse the returns. Either to fight fires or to light them requires a quick concentration of attention and crews, and at least a temporary surplus of money and grassy fuels.

Yet this is exactly what the western plains typically lack. Even that keystone agency of American wildland fire, the U.S. Forest Service, must struggle to overcome those powerful pressures toward institutional entropy.

The Nebraska National Forest is a curious, and in many respects, contrived artifact. It has the largest afforestation project in the United States, patches of mixed-prairie grasslands, and a slice of western woodlands, including a wilderness. They have little in common—not geology, not ecology, not history, only their assigned jurisdiction and collective name. Just that seeming dissipation, however, makes them collectively into a cameo of fire management on the plains.[2]

"The Nebraska," as it is usually called, began in 1902 when Charles Bessey convinced President Theodore Roosevelt to proclaim two as-yet-unpatented chunks of the Sandhills as forest preserves with the intent of planting them to pine. In the 1930s the federal government acquired an archipelago of abandoned Great Plains ranches, declared them national grasslands (by analogy to national forests), and assigned them to the Forest Service for administration; the Nebraska gained three such units: the Oglala, the Fort Pierre, and the Buffalo Gap, the latter of which encircles Badlands National Park. The 1950s added a strip of Pine Ridge, a rudely forested remnant plateau, like a barrier reef outside the Black Hills, out of which, in 1986, some 7,794 acres were designated as the Soldier Creek Wilderness.

A map of its parts looks as though a stale cookie had been dropped on a floor and broken. From forest headquarters at Chadron to the Fort Pierre District is 238 miles away by road. From Chadron to the Bessey Unit is 194 miles and over three hours distant. The reality, in fact, is worse. Most of the grassland units and the Pine Ridge District are a coarse sieve of private and public holdings. A detailed cartography of ownership would render a landscape that resembled a prairie dog town full of holes and bare patches. The Nebraska has 5 fire-management units, 12 special interest areas, and 5 research natural areas, none contiguous. With minor exceptions its borders trace old land survey lines, not geomorphic features. The "forest" is a scatter diagram of land use, and how to draw a regression line called fire management through those points defies any easy administrative algorithm.

Wildfires come often—dry lightning is common. But they are not frequent enough or big enough to warrant a large on-forest suppression organization. Fires blow up and blow out within one or two burning periods. By the time an engine could drive from Chadron to Draper, a typical fire would have expired. Initial attack is largely done in cooperation with VFDs, most of which are not NWCG qualified. Once or twice a decade a fire persists long enough to involve serious extended attack and a call-up of outside forces. When that happens the incoming and the indigenous often clash. The cost of building up forest capabilities to meet the once-a-decade burn is too high, but so is the cost (both fiscal and social) of upgrading capacity among cooperators. The fires will happen. Some will get big too quickly to do more than wait and contain them after they've made their run. In July 1989, a month after its formal designation as wilderness, the Fort Robinson fire bolted over the Soldier Creek Wilderness in an afternoon and killed some 90 percent of its resident forest.[3]

Circumstances are no less subversive of prescribed fire. There is no lack of interest. The Bessey District wants to burn to drive off eastern red cedar. The Pine Ridge District would like to underburn the pines to reduce fuel and, locally, to protect Chadron's watershed. The Fort Pierre District is experimenting with patch-burning for grazing. Chadron State Park wants to thin out woods and control cool-season invasives. The difficulties are, again, mustering capacity to do the burning on the short-notice windows that open and close with the same suddenness that sustains wildfire, and, ironically, with stubborn questions over fuel management.

The National Fire Plan and its successors have allocated considerable funds for fuels, but they are fuels projects designed to reduce the threat of catastrophic wildfire and only indirectly to support prescribed fire, and then in the service of quelling conflagrations and protecting the wildland-urban interface (WUI). That money has shrunk dramatically, and most of it in the Nebraska, as at the Black Hills, goes to coping with the fuels generated by a plague of mountain pine beetles or with silvicultural pruning—thinning—at the Bessey. There is little left for grasslands, which constitute the bulk of the Nebraska's estate. Worse, the fuels problem on the prairie is to stockpile combustibles, not reduce them. Paradoxically this costs money since it requires reducing or removing livestock, which means buying out permittees (the obverse of paying logging firms to haul off small-diameter timber). It gets worse yet. The reason for the burning is not to prevent blowups but to regenerate prairie. Ecological integrity is not, however, the stated mandate of fuels programs.

A first glance suggests the issue is scale—a reading that slots nicely into a reliance on geographic information system data banks and current mantras about fire knowing no borders. The solution is cooperation, for only an alliance of partners can respond to wildfires that rise and fall like spring blizzards or to prescribed fires that demand sudden roundups of engines and torches. A further review, however, points to more elemental factors.

The conundrum at the Nebraska is not simply scale, the ability to reach beyond district resources for either emergency or opportunity. It is the capacity to reach, when needed, a critical mass—of fire crews, of funds, of grass, of organization, or to put it differently, a surplus adequate to overcome the tendency of human enterprise on the plains to diffuse or dissipate outright. What is striking about the Nebraska story is that even a federal presence is not enough by itself. The Nebraska National Forest and Grassland can hold the pieces under a common roof. It cannot alone manage them from a common point.

Historically, plains peoples found ways to live with that tendency to disperse by keeping on the move. They applied their small numbers with force at certain times and places rather than try to hold the whole at once. Later American settlement, after a phase that tried fixed settlement in sometimes brutal isolation, relied on transient laboring gangs and machinery to move cattle and harvest grain, and on banks or government to tide over through drought, deluge, and locusts. No one could succeed without appeal to some outside supplement, but that supplement had to work with, not against, the diffusionary instinct of the plains.

If the nation's most bulky fire agency cannot solve the problem, it's unlikely that any of the other players can either. The states haven't the resources, have little fire capabilities of their own, and regard NWCG standards as beyond their means. Nor can private landowners, even fire-adept nongovernmental organizations (NGOs) like the Nature Conservancy. What is emerging is a dual strategy in which an outside entity acts as a catalyst or supplement and the system that results relies on its capacity to move.

An impressive array of institutions is capable of supplying the critical extra that local communities need but can't afford. The major federal fire agencies of course will sign cooperative agreements. The Natural Resources Conservation Service can assist with advice and select grants. The Conservation Reserve Program (CRP) banks grass from farmland. Pheasants Forever will furnish a fully equipped burn trailer to formal burn associations. Prescribed fire councils offer model organization bylaws. FEMA will help fund VFDs, a handful of which have begun to hire out for prescribed burning. And there is plenty of intellectual firepower from universities, federal research, and extension services. If a community needs help, it can get it.

What is distinctive about the northern plains (to keep the Nebraska in focus) is an emerging reliance on seasonal crews who can move as needed. The Nebraska oversees nine discontinuous parcels scattered across two states. Wind Cave National Park in the Black Hills oversees seven park units. The Fish and Wildlife Service sends burning crews around the northern plains. The Nature Conservancy holds training exchanges, in which it offers prescribed fire qualifications to agency personnel on sites it wants burned. On some sites contractors—the contemporary version of migratory shearers or sheavers—do the work. The effective model is not simply one in which neighbors coalesce as needed (as, for example, mutual aid by VFDs or prescribed burn associations). It is one in which fire managers overcome distances of space and time by moving.

O. E. Rölvaag set his famous novel of settler life, *Giants in the Earth*, north of Nebraska. He surely intended his observation about how hard it was "for the eye to wander from sky line to sky line, year in and year out, without finding a resting place" to apply to those unsettled minds yearning for reference and roots. The traditional solution—to migrate with the seasons—flew in the face of colonization based on stable blocks of land under single ownership; Rölvaag's Norwegians had already crossed an ocean and half a continent and did not wish to adopt wandering as a life, even if two generations later many of their descendants emulated the migratory fowl and became snowbirds.

But his sentiment might also characterize a traditional fire management that sought fenced practices and rooted norms. After all, the Nebraska began by planting a forest where one did not exist and sought to stabilize sand that blew with the winds and burned when its grasses cured. To succeed today, however, the Nebraska requires a different understanding and must find ways to work with, not against, that restless horizon. Such a fire program may sound counterintuitive to expectations. It will likely emphasize grasses, will seek to build up fuels, will organize fire management not on fixed lookouts and fire caches but on the ability to move with the melting snow and gusts through tufts of little bluestem.

LOESS HILLS

WHEN THE BROHMAN patriarch bought the family ranch in 1943, the only trees on it were two native junipers planted outside the kitchen. The Loess Hills of Custer County were slick with grass. Today, what had been a mixed-grass landscape is fast becoming a monoculture woodland of eastern red cedar salted with occasional clumps of hardwoods. The region's fires have shape-shifted in a comparable way.[1]

Once fires had flowed as insouciantly as water until a century of settlement radically restructured both. The hydrologic regime bifurcated into floods and irrigation. The fire regime split into wildfires and prescribed burns. But unlike its waters, the region's fires did not helpfully flow in channels or reside underground, from where they could be pumped to the surface on demand; they free-burned where wind and fuel permitted, and they could only be managed on a landscape scale. That requirement, however, is exactly what recent history has frustrated. The relative unity that glaciation had imposed, American settlement has fragmented.

There are in reality three overlying landscapes. One is the tumbling vista of grass and patchy woods that stretches to the horizon and could, in principle, accommodate recurring fire as it had for thousands of years. The second is a crazy quilt of owners, institutions, and purposes that made fire exclusion and wildfire the default setting. The third is the historical landscape that explains how the system flipped, like an iceberg overturning, from a diverse prairie to a woodland encrusting the hills like an ecological scab.

The surprise is not that, in the absence of fire, trees overwhelmed grasses in semihumid grasslands, which is a story as old as settlement on the Great Plains. Nor that garden-variety ironies and paradoxes grow rank as ragweed; Nebraska originated Arbor Day in 1872, and the cedars were specially selected by arborists to flourish more tenaciously than the native junipers, and they were then promoted by officials as windbreaks and amenities before they snapped their leash and overran the landscape.

What catches the mind is how rapidly the cedar can take over and how onerous is the task of reversing the epidemic once it has established itself. Each landscape has, in effect, its own prescription.

The Lower Loup Natural Resource District embraces five million acres of loess soils, of which three million are grassed. The hills are ranched, and the broad floodplains along with level valleys and mesas are farmed. The lowlands, spanning both sides of the 100th meridian, often require supplemental irrigation by canal or pivot. Many of the hills are too steep for plowing without serious erosion. As settlement progressed, a low-grade transhumance evolved, in which herds moved into the hills during the summer and fed on cropland residue and silage over the winter.

It's the kind of seasonal migration that typically encourages anthropogenic fire. But there is little evidence for systematic burning. What burning occurred came from wildfires, which were sufficiently ample to keep woody niches in check, but which also terrified the populace, who kept burlap sacks and creamery jars full of water on their porch, ever ready to attack the feared smoke on the horizon, and who relied on intensive grazing to hold the grassy fuel in check. In time, the system shed its mobility: farmers farmed, and ranchers ranched. By the last decades of the 20th century the hills were being slowly emptied of people, and many of those who bought into the land were absentee owners who leased out to herders. Throughout, the fear of fire persisted, while techniques for keeping it out or finding internal-combustion surrogates strengthened. Fire became rare, and as it faded, so did the prairie.

Surely, woody shrubs and trees had always existed—had rooted in the soil as soon as the glaciers ebbed. Over time the dominant species changed, and as the plains warmed and burned, and grasses dominated, they had retreated to refugia such as wet bottoms, rocky outcrops, or the lee side of meandering creeks. Then the native browsers that had trimmed them and the fires that had swept away seedlings disappeared. They could spread farther afield. The real shift, however, arrived when tree planting came into vogue to ameliorate climate, when shelterbelts began to run over hills like hedgerows, and when a hardier species of juniper was promoted by the powers that be. The performance-enhanced cedar went wild, taking hold in steeper canyons and on north-facing slopes before spilling out across the hills. The copses made roundups tricky as cattle (especially bulls) hid, untouchable, in thickets. At first the wooded patches went unnoticed, or added a pleasing limn to the rounded landscape, furnishing a visual trim. They expanded, however, at near exponential rates. One year they were in the crevasses and bottoms; the next, they pocked a hillside like mushrooms; a few more and they crowded out the grasses. The effective pasture shrank. The whole habitat sank under the weight of the cedars.

The decay was worse because the cedar suppressed the grasses that powered the culling fires. That new fire regime—a regimen of burning only by rare and random wildfire—permitted exotics like cool-season Kentucky bluegrass and early maturing smooth brome to establish and outcompete the indigenous grasses and to sop up the soil moisture that the native warm-season grasses and forbs needed. As the isolated trees thickened into patches, and patches into groves, their roots drained away that moisture still deeper. Some exotics (think buffel grass) inflate fire; others diminish it. On the Loess Hills everything seemed to push the landscape away from routine surface fires and into a boom-and-bust cycle in which nothing burned or everything did.

For a long time, too, ranchers kept the same stocking rates, unaware of how much pasture had been lost, which only worsened the overgrazing, which made a cleansing fire even more unlikely. When fires did occur, they were effective only in checking the very young reproduction. Once a tree matured and thickened, or tree canopies locked, one to another, the grass beneath disappeared, and flames blew against and around the canopies, or only scorched the lower branches, which rendered the cedar even more impervious to future fires. Once the cedar had established itself, only the most intense burn could uproot it. It was an old and tiresome Great Plains story: a brush infestation that converted what the plow had spared. Prairie fire disappeared because the prairie did.

It was possible to cut or bulldoze off the cedar, but that only broadcast the seed ("as though I'd planted the hillside with a drill," one rancher wryly noted). It was possible to kill cedar with herbicide, but that was expensive and left seed on the ground ready to sprout. But without fire there was no economical way to scrape the grassland clean. So ranchers cut and piled, and they cut and stacked against the thickets, hoping to stoke a fire fierce enough to gnaw through the canopies. The best burning conditions, however, cast firebrands upward into plumes and made escapes common. Unless the land was burned when the cedars were seedlings the infestation would return, and unless adjacent lands were treated reinfection would occur immediately.

An ecosystem built on a suite of prairie grasses and forbs, its pieces soft-welded by frequent burning, fell apart, piece by piece. The shift made fire more difficult, or when the cedar could burn, it sparked explosive outbreaks that erupted into plumes and blew sparks like a dust storm. The longer fire remained out of the system, the more difficult good fire became, and the more probable bad fire. Too often when ranchers, in desperation, tried to burn they executed it poorly, which only enhanced skepticism. In 2011 three amateurs died in a botched burn at Trenton that caught the attention of the state legislature and threatened to dampen burning further in the name of public safety.

No landowner, even with a driptorch, or a dozer and torch, could fight off the infestation alone. Eastern red cedar had propagated promiscuously because fire exclusion had operated on a landscape scale. Fire's restoration could only succeed on the same terms.

The locals needed leverage. They needed to burn hotter, which meant larger patches—thousands of acres—to catch potential slopovers. They needed to burn across the surveyed grid of townships. They had to designate borders that made sense from fire-behavior considerations, which meant it would involve many landowners. Small patches were hard to hold and subject to constraints imposed by many neighbors. It was a matter of scale.

Leverage also meant social power, a counterforce against the depopulation of the interior hills or its conversion to absentee owners and lessee herders. Eradication meant landowners agreeing to cooperate, workers willing and trained to do the burning, expensive equipment beyond the reach of any one rancher, advice about how to overcome startup costs. Support came—had to come—from the outside. The Natural Resources Conservation Service (and its predecessor, the Soil Conservation Service) could assist with advice and some funding to help clear and burn. The Conservation Reserve Program allowed land to lie fallow for a decade until a burn brought it back into production—a stockpiling of combustibles and experience. The Nature Conservancy hosted Fire Learning Networks in which burners acquired and renewed their qualifications by helping burn. Pheasants Forever boldly contributed a prescribed burning trailer stocked with personal protective equipment, flappers, driptorches, and radios to any formal prescribed fire associations, along with access to liability insurance.

Those associations—voluntary collectives of locals who agreed to help one another burn—began to appear. The original settlers had successfully fought back prairie wildfire only by turning out as a community and, with what often seemed to them terrifying slowness, by converting field after adjacent field to incombustibility. Sod busting was fire busting. The future of fire management in the Loess Hills required a comparable effort, for the spread of eastern red cedar resembled an epidemic, and prophylaxis required wholesale inoculation, or else the unvaccinated sites would continually reinfect all the others. Private actions from individuals acting on their own land and in their own interests had broken the old regime. Only a collective response could pick up the pieces. In 2012 the Custer County Prescribed Burning Association completed its first year by sponsoring a dozen burns for its 40 members.

Without fire the future of the Loess Hills is clear. The land will so overgrow with cedar and miscellaneous hardwoods that the proverbial squirrel could travel from the Loup River to the Sandhills by leaping from tree limb to tree limb. Grasslands will exist only as lawns, an ironically inverted image to the shade trees originally planted around the farmhouse. To keep pasture, landowners will have to mechanically slash and carry, and repeat the clearing every 10 to 15 years. The expense will make ranching suitable only for hobbyists or for parks and preserves keen to maintain a semblance of history and nature.

But paradoxically the eastern red cedar, by both retarding fire and demanding its use, may provide a point of positive infection for landscape burning in the name of ecological benefits. Once the mechanism for burning is established, it could move from trees to pastures, and from the eradication of cedar to the control of exotic grasses and the promotion of habitat favorable to pheasants and other creatures that once gathered on the prairie. Once the initial rotation occurs, the improvement in forage makes up for the down payment in grass needed to burn. And other subsidies exist: the Environmental Quality Incentive Program under the Natural Resource Conservation Service, and the Conservation Reserve Program. Returning land to cultivation with a burn helps create fuelbreaks and fosters familiarity with burning. Such programs overcome what stalls many experiments: the need to stockpile grass (which is to say, forage) by keeping slow combustion out long enough to support fast combustion.

For decades—save at the Flint Hills, always an exception—what interested researchers and practitioners on the Northern Plains was fire in the service of prairie restoration and biodiversity. In such places managers have recognized fire as a keystone process both natural and necessary. What makes the Loess Hills special is that efforts here aim to restore fire to working landscapes—not celebrity sites, not pristine preserves, not gated communities for endangered species. In the Loess Hills fire is asked to do what it has always done for humanity: to make the world more habitable.

It's an easy case to make in principle. Putting it on the ground is where the gears jam, the accidents happen, and the years creep by. In the Loess Hills it's happening acre by acre, person by person.

It had been a busy spring. An exceptionally warm and dry winter had advanced the calendar, shrunk the season for burning, and primed the countryside for explosive fires. John Ortmann of the Natural Resources Conservation Service on the Lower Loup District found himself constantly in the field—this after a winter (actually, several winters) of intensive planning, and after a long and wending career.

In 1978 he graduated with a degree in journalism, which 18 years later had him writing copy for an advertising agency. It wasn't enough, and he reentered the University of Nebraska–Lincoln for a BS in ecology, which segued into an MS in 1995, and then a PhD in 1998. His research interests in eastern red cedar reflected those of his advisor. Upon graduating he took a low-paying position at Colorado State University in the extension services program, where he learned to deal with private landowners, small acres, and the need to translate academic learning into practice. Then the soft money dried up. He accepted a position at the Nature Conservancy's Niobrara Valley Preserve, where he lived on site. He helped initiate a series of national prescribed-fire training exchanges. After a falling out, he took a position with the Natural Resource Conservation Service in Ord, Nebraska.

If it was an odd career trajectory into fire management, it was no odder than the fire manager himself. Ortmann had hemophilia. The disorder attacked his cartilage and other tissues where and whenever internal bleeding had occurred, and as a result of some childhood injuries it had trashed his ankles and elbows. Walking was painful, more hobble than stride. To end up as a fire manager made a curious outcome for such an inheritance, but if you study *Juniperus virginiana*, you will come to fire, even if you approach it through hemophilia rather than hotshotting.

The 2012 season started in January with five burns spread over three weeks. It had been a mild winter, but the loess soils were impassable—slick as ice—until frost had left the upper crust. The window of opportunity opened between the time the soil was dry and the trees began green up. Most years that meant March. Moreover the International Migratory Bird Treaty demanded burning end by April 1 so it would not interfere with nesting. In March Ortmann, local ranchers, the Nebraska Natural Legacy Project, the Nature Conservancy, the Loess Canyons Rangeland Alliance, the Eastern Loess Canyons Rangeland Alliance, the Custer County Prescribed Fire Association, and Pheasants Forever, assisted by trainees from the national exchange program, began burning selected sites. In April he moved into CRP lands, which allowed exemptions to the bird treaty, at least until May 1. Wherever possible they burned on a landscape scale, which meant setting perimeters broadly and in ways that would allow fire to free-burn, which demanded cooperation among perhaps a score of landowners. Many projects never got beyond the planning stage.

The size of the patches ranged from several thousand acres to as tiny as a single acre. Fire behavior trod along a prescription ridgeline between not burning at all because grass or foliage had greened up and, during drought, exploding through closed-canopy cedar groves. On April 14, having sat up to 3 a.m. watching cedars mass torch on the Willow Creek unit, Ortmann collapsed from exhaustion on his driveway. On the 24th, while trying to mop up the interior of 520 acres left over from the Training Exchange Complex, his ATV tipped over on an eroded cow path. He threw himself clear and then used his legs to hold the vehicle from rolling over him. Gas dripped from the ATV, while a small spot ignition glowed a few inches from it. He was pinned. Slowly, he was able to push the vehicle upright onto the cow path and then quickly extinguish the flame. Embedded within a ravine, his radio would not work. He blacklined around the ATV and walked to the hilltop to raise help. Two days later he was rushing to complete the final four burns for which he had plans before the May 1 deadline shut burning down for the season.[2]

Not your typical fire career, nor a typical prescribed burn, nor the kind of landscapes or institutional arrangements that enter textbooks or Lessons Learned bulletins or find their way into the endless memoirs inspired by *Young Men and Fire*. But the Loess Hills are not wildland: they are rural and private. Fire management did not evolve here out of fire suppression, which is left to VFDs; rather, it emphasizes planning and burning. In place of Type 6 engines and helicopters, burners rely on tractors, ATVs, and cattle sprayers. Fireline operations are organized but not according to the incident command system. Interagency cooperation is paramount but it does not occur among government-sponsored agencies. The landscape is one of NGOs and landowners, with public assistance for some funding and guidance. The program is cobbled together and run on a shoestring. But fire management is just as risky, and just as essential.

The Loess Hills of Nebraska encompass a biogeographic province twice the size of Yellowstone National Park. The 19 burns and 4,682 acres fired in 2012 exceed the output of all but three national parks in the preceding year.

KONZA

TALLGRASS PRAIRIE is to the Great Plains what the longleaf forest is to the coastal plains, and big bluestem is as much a signature species as longleaf pine. Both biomes need fire. Both were so decimated by settlement that only 2 to 4 percent of the original biomes remain extant. Both have been the subject of intensive preservation efforts. Both have served as a poster child for fire science. Any survey of the Southeast—even the ecological equivalent of a planetary flyby—will include longleaf, particularly such residual patches as the Wade Forest in the Red Hills of northern Florida; and any reconnaissance of the Great Plains will find its orbit deflected by the gravitational mass of Konza Prairie in the Flint Hills.[1]

The northern flanks of the Flint Hills sit close—100 miles—to the geographic center of the continental United States. So, too, they crowd near the core of tallgrass. The Flint Hills were too rough to plow and too remote to pave. They survived the surge of agricultural settlement that elsewhere ground over the plains and that here turned instead to ranching. In the post–Civil War years the hills became an open feed lot from May to June for Texas cattle being shipped to Kansas City. The ranchers burned to freshen the glorious native grasses; they fired every spring, typically in April. That regimen put the hills in production and kept them burned. The prairie thrived. The Flint Hills had more tallgrass than anywhere else. They were to tallgrass what the southern Sierra Nevada was to sequoia groves. When the time came to protect that fast-dwindling habitat, the hills gathered into one place the whole tribe of reasons for doing so.[2]

There were many purposes behind prairie preservation, and repurposed ranches furnished the means and scale necessary. Because the land was grazed (and burned) and not plowed under (or its fires extinguished), the basics of the original biotic matrix survived. Ranching persisted, and burning with it, most spectacularly along the hills' blurred terraces. Both Tallgrass Prairie National Preserves are here—the National Park Service's outside Strong City, Kansas, and the Nature Conservancy's in Osage County, Oklahoma. Konza Prairie Biological Station triangulates the northern lobe of the hills, south of Manhattan, Kansas.

Few preserved prairies lack a research component, targeting whatever special feature of the landscape led to its being set aside. Some survive because, as with ranching, they had an economic rationale. Some were created for historical and cultural reasons—the restoration and preserved memory of presettlement conditions. Others promote biodiversity or tourism; there are even pay-to-burn festivals in which visitors can experience the thrill of setting the prairie aflame. Ranchers want fire research to help boost productivity. Preservationists support science to better understand how to protect extant patches and rebuild prairie out of cornfields. The tourist industry wants visitors to experience *Little House on the Prairie*, or the wonder of cross-continental trekkers on their way to Oregon, California, or Utah, and they need fire to sustain something of that original scene. Biodiversity interests, too, yearn to know what regimen of fire or of fire in association with other processes best promotes particular species or biomes. Each looks to research to create data and turn that data into practice. But only Konza is dedicated to full-spectrum, basic science.

Its 3,487 acres are managed as intensively as any commercial farm. Konza is not a natural wilderness, left to its own devices. It is not a tourist destination, open to hikes, picnics, and recreation. It is not committed to maximizing commodity production. It is not outfitted with gift shops, simulacra of sod homes, or faux pioneering museums. It's a field station set up in 1971 and run by the Biology Department at Kansas State University; a long-term ecological research site under the auspices of the National Science Foundation since 1980; and the primary facility for understanding the dynamics of an intriguing, deceptively complex, and vulnerable ecosystem nearly as old as the Holocene and threatened by the cumulative abuses of the Anthropocene. The research agenda at Konza studies every aspect of the biome. Not least, it is one of the few research enterprises anywhere created from its origins to study fire. It is to tallgrass what Tall Timbers Research Station is to longleaf.

If the Flint Hills have become the homeland of tallgrass prairie, Konza has emerged as the founding lab for prairie science. It provides the strong nuclear force of data and concepts that hold collective understanding together.

Why Konza? Why fire? The story opens, like a door, on two hinges. One is the recognition in the postwar era that tallgrass might disappear if it was not actively protected and that its disappearance would hurt science, along with other prairie values; the need existed for a field station dedicated to research on land that more or less retained its critical properties and processes. The other is the personality of the project's prime mover, Lloyd Hulbert.

Observers early recognized the association between prairie and fire, but they disagreed over whether the fires were an epiphenomenon that just happened or whether they were integral or even informing, without which the biome would vanish. They knew that if fire were removed, brush and woods would overrun the grasses, but they did not know if fire as biocombustion was necessary or if some other, mechanical means to clean out the thatch and repel hardwoods would work as well. They could agree with Aldo Leopold as he eyed the Wisconsin prairie and noted the "ancient" ally the grasses had in fire as a flaming brush cutter to beat back burr oak. The Curtis prairie at the University of Wisconsin Arboretum decided in the late 1940s that it had to include fire in its efforts to return farmland to prairie. The first prescribed burn by the Nature Conservancy occurred in the Helen Allison Savannah outside Minneapolis in 1962. And while everyone could see that tallgrass thrived amid the annual burning that lit up the Flint Hills every spring, they did not know how those grasses might flourish if fire were removed. They lacked the disciplined knowledge that could only come from controlled experiments. They burned because burning worked—had always worked. They relied on tradition, folklore, and operational pragmatism.

Something else, however, had to tweak that perception into a research agenda and to distinguish Konza from other places that studied grasslands. What happened was Lloyd Hulbert, the genius locus of Konza, who for several years during World War II led an improbable life as a smokejumper in the Northern Rockies. Fire did for him what it did for the big bluestem that he preferentially studied.

Hulbert was born in Lapeer, Michigan, in 1919 and raised a Quaker. He earned a degree in wildlife conservation from Michigan State in 1940. When the United States entered World War II, he signed up for Civilian Public Service as an alternative to the military draft. He spent a year and a half in a camp in Michigan, and then, in one of fate's happy quirks, the program offered a chance to serve in the fledgling smokejumper corps, which was just going operational when Japan bombed Pearl Harbor. He spent the next two years in Missoula, jumping on Northern Rockies fires and, during the off-season, doing rangeland reseeding through the Intermountain Experiment Station. He was discharged in April 1946.[3]

For Hulbert the experience meant a sudden immersion into a very different life—mountains instead of plains, sweeping forest fires instead of field burns in stubble, practice rather than research, all amid a daring occupation barely beyond its beta phase. If only by osmosis, he encountered a distinctive fire culture. It was one committed to fire's suppression, but it put fire at the core of life and understanding. His rangeland labors for the Forest Service, though not all of it specifically in fire, still put him in an organizational culture for which fire was an enduring obsession as well as a bureaucracy whose research mission was tent pegged by a network of experimental forests and ranges.

For the next few years he taught as an instructor at Montana State and Minnesota. In 1953 he went to the State College of Washington (now Washington State) to study under Raymond Daubenmire, the doyen of grass biology, where he worked on *Bromus*, another pyrophyte. (He's credited with working out the extraordinary root system of cheatgrass that makes it so effective an invasive.) In 1955 he joined the faculty at Kansas State University. Here he found himself proximate to another fire culture, this one tethered to the Flint Hills and working under a dean, Frank Gates, especially keen on academic field research stations, and a department that had considered one since 1953. He plunged into research on prairie grasses and grass-tree interactions, initially emphasizing soils, then fire.[4]

Over the next decade Hulbert campaigned for a network of protected areas in the state that could also serve as control sites for research; the notion was broadcast as "A Plan for Natural Areas in Kansas" by the Conservation Committee (then under Hulbert's leadership) of the Kansas Academy of Science and enacted as legislation in 1974. The crown jewel was the Konza Prairie Research Natural Area six miles south of Manhattan, which served as the Kansas State Biological Field Station. With funds from the Nature Conservancy, the ranches that became Konza were acquired from 1971 to 1977. Former ranch buildings morphed into labs and housing for visiting researchers. Burning for pasture evolved into systematic experiments of fire regimes. In 1987 Konza acquired a herd of bison to help round out its complement of grazers. It lacks a similar suite of browsers, and its scale is too small to accept the wolves and bears that were natural predators, to say nothing of human tribes, so scientists must serve instead to cull the herd and burn according to the prescriptions of experiments rather than the imperatives of hunting. As fires kindled by aboriginal firesticks had ceded primacy to wooden matches and ropes soaked in kerosene, so now they yield to the strike of a steel will on the flint of the hills.[5]

Since its founding, Konza has enjoyed a steady drumbeat of expanded ambitions and recognition. It was one of the six founding sites for the National Science Foundation's Long-Term Ecological Research program. It became a research site for NASA, for the USGS (National Benchmark Hydrologic Network), for UNESCO's Biosphere Reserve project, and for the National Ecological Observation Network. In a couple of decades it had gone from working ranch to world-class research center, part of a global consortium committed to understanding grasslands. Not everything known about tallgrass prairie comes from Konza, but no synopsis can ignore it, and it effectively synthesizes all the others.

When it began, Konza was not the only academic field station in the United States or even the only one dedicated to prairie. Agronomists fretted over fire and range, and agricultural scientists at Kansas State (and elsewhere) were burning plots by 1926. In 1933 the University of Wisconsin had acquired 60 acres of abandoned land for its arboretum and began to reclaim it, with some assistance from the Civilian Conservation Corps. A decade later, under the leadership of John Curtis, it commenced formal experiments in restoring pasture to original prairie; some experiments included fire. The project was arduous and complex since the land had been plowed and farmed, then left for fallow, and finally used to pasture horses. Restoration from the gashes, feral cultigens, and the camp-follower weeds of settlement remained its raison d'être.

Konza sprang from another vision. Hulbert wanted a large site capable of multiple, long-term experiments on land that, while grazed by cattle, which grazed differently than pronghorn and bison, had remained prairie throughout. Unlike other experimental grasslands, this one would investigate the full gamut of prairie processes, would have a scale that made it more ranch than kitchen garden, and would include fire as a founding feature. Lloyd Hulbert never penned his intellectual autobiography or traced a genealogy from Rocky Mountain snag fires to the rituals of Flint Hills burning, but something had to pique his curiosity that did not happen to others, and it is likely that his tour as a smokejumper made fire, as fire, a topic for him worth pursuing. Certainly those who knew him thought so. Uniquely, Konza embraced both prairie research that involved fire and fire research that occurred on prairie.[6]

Today, Konza organizes its research around the three themes that most inform tallgrass: climate, grazing, and fire, or, what really matters, how they interact. In the spring of 2014 there were 165 active research projects on what is probably the most heavily instrumented patch of prairie in the world.[7]

The lesson that repeated, like a fractal, at all scales, is that one effect by itself matters less than the ways it interacts with others. To measure those effects requires a large landscape to accommodate at least some of those exchanges and their experimental manipulation. The separate parts get simplified into dichotomies—soil shallow or deep, bison grazing or not grazing, herbivorous insects or their absence, burned or unburned, drought or deluge. That is how reductionist science (and the human mind) tend to work. But those factors do not behave in the field like toggles that simply switch on or off: they are rheostats that modulate through a range of conditions. Some sites might be burned annually, and others once a decade; some patches are grazed routinely, others infrequently. The permutations among all the possible interactions probably equal the number of atoms that make up Earth. No science can test for them all, but if Konza cannot replicate the world's "blooming, buzzing confusion" (as William James called it) with controls, it can at least capture some of its richness with studies of second- or third-order complexity. Ideas, too, achieve their power from their interactions.

The results can be counterintuitive. What matters are the linked regimes of rain, grazing, and fire, the three-legged stool of prairie ecology. The right mix can enhance biodiversity and ecological stability. Fire doesn't control sumac, for instance, unless the burn is really hot and probably comes midsummer, and it still needs browsers to work a one-two punch to drive the brush back. Herbivorous insects stimulate more budding, and the bud bank, not the seed bank, responds best after fire. Bison and burning, together in the right mix, increase forbs and dampen the trend toward monocultural grasses. And so it goes, fractal after fractal, from bison to grasshoppers, from flowering bluestem to buds, from landscape to patch. Grasslands offer opportunities for managed manipulation on an annual basis not available to forests or brush. And Konza offers the scale in space and time necessary to grasp a bit of the complexity, even if its insights seem little more than fireflies captured in the gloaming.

Its location allows it to investigate fire-grazing dynamics at all levels and scales. Its size—almost 60 times that of Curtis—means it can accommodate comparable, controlled experiments at a watershed level. Its commitment to the long view means it can track secular changes in climate and brush encroachment, and that it can test the possibility of ecological reversibility. Can a system that had remained unburned for 20 years be "restored," and if so, by what combinations of fire, grazing, and climate? It can imagine fire as an ecological catalyst, as a synergistic process, as something embedded within a biological matrix. Its grasses are not simply fuel; they are also feed and fodder, and the rush and rhythms of biotic actors determine the character of the fires as much as wind and drought. These are unique features of the Konza infrastructure and why the National Science Foundation has renewed its status as a Long-Term Ecological Research site for the seventh time.

For Lloyd Hulbert it was the task of a lifetime. Honors eventually came, among them the Nature Conservancy's Oak Leaf and President's

Stewardship awards. But Konza as a working lab was his real testimonial. There he personified the old adage that a single spark can start a prairie fire. In truth, he started hundreds, all set within a meticulous matrix of experimental designs and imagined over the span of a lifetime and beyond. For Hulbert the long view was always the better choice. In September 1985 he wrote his sons that "most people try to ignore the subject [of death], which is the wrong thing to do. Death is necessary and sure for everyone. If there were no death, there could be no birth." Eight months later his own death came. His ashes were scattered over Konza's bluestems.[8]

PATCH-BURNING

WHERE PRAIRIE MEETS WOODS is a biotic border that spans a continent, and it displays a continent-sized roughness. In rude terms it traces the coarse shoreline between a sea of grass to the west and a land of mixed forests to the east, an edge sculpted into the ecological equivalent of bays, narrows, skerries, and estuaries, as climatic tides, the tectonic lurching of glaciers, and the sprawl of colonizing species have tugged and twisted, and here and there allowed grass or woods to mostly prevail. That textured shoreline holds a jumbled geography of incombustible wetlands and free-burning bottomlands, fire-flushed barrens and fire-hardened forests, prairie peninsulas and prairie patches, oak mottes and woody copses, and landscapes latent with bits of them all, some extending over hundreds of miles.[1]

It is a fractal frontier, patchy at every scale, with small patches within larger. And it is a frontier of fire, with each part checked or boosted by the ferocity and abundance of burning.

CROSS TIMBERS

Even so, the Cross Timbers stand out. They proclaim a bold, woody headland, as distinctive as the White Cliffs of Dover, between the grassy sea that swells to the west and the humid forest that crowds the east. It is here that storm surges of fire, roaring over the long fetch of the

Great Plains, whipped by the westerlies into whitecaps of flame, crash against the less combustible woods. The belt is long, stretching from the Edwards Plateau of Texas to the Flint Hills of Kansas; irregular and sinuous, roughly cruciform, historically varying from 5 to 30 miles wide, but at places spanning most of Oklahoma; and persistent, its 11.5 million acres defying settlement's attempts to log, plow, graze, or burn it into oblivion. Instead, it continues in Oklahoma to thicken with stubborn oaks—blackjack, shin, live, and post.

The contours of the Cross Timbers roughly track soils, a divide between grass-promoting limestone and the oak-favored sandstone. But they also trace a kind of biotic dry line, jumping west to east from 26 inches of rainfall annually to 42 inches. To the east, ahead of fronts, moisture surges up from the Gulf of Mexico and brings rainfall sufficient to sustain woodlands. To the west, weather systems draft air from the deserts of northern Mexico and west Texas; there is less moisture, and it promotes a regime suitable for shortgrass prairie that leaves its woods strung along streams as gallery forests. It is the middle ground, the belt of tallgrass prairies and implacable oaks, where the most vigorous fires meet the stiffest woods, a kind of tornado alley for flame. Only the most savage fire top kills the dominant trees: mostly the oaks freshly sprout, hydra-like, into an enduring, oft-impenetrable thicket, a living seawall. Even the wildest fire surges break against it. Washington Irving famously described the outcome as a "cast iron" forest. He didn't mean the phrase as a compliment.

But he might have. The Cross Timbers endure, a patch of history as much as of geography. They remain the dominant ecotype of Oklahoma. And where they intercalate with prairie and city, they display the peculiar patch dynamics of a unique American fire regime. Each patch has its own internal regimes, but it is how they all link that defines combustion across the region today.

PRAIRIE PATCH

The Flint Hills parallel and intercalate with the Cross Timbers, stretching from northern Oklahoma into southern Nebraska. As their name suggests, they were too rocky to be plowed, so they became a site for ranching. Elsewhere prairie shattered because sedentary settlement through roads, fields, and towns broke the power of fire to propagate. Relic patches remained in odd niches, such as along the burned right-of-ways of railways or in places where terrain frustrated plow and grader, and the land remained in grass.

The prairie patch within Osage County claims the southernmost reach of the Flint Hills, and it has survived more or less intact precisely because it is both grazed and burned. The linkage is deliberate: it is burned to improve grazing, and because it is grazed it gets burned. Because it gets burned as part of an annual routine the greater prairie patch displays a fire culture that has long disappeared from America's vernacular landscapes. That tradition has kept fires that have elsewhere vanished.

These are working landscapes. Ranchers seek to maximize their economic return and use fire because it assists a pattern of raising cattle. The norm is to burn early in the spring to help kindle a burst of warm-season grasses. They burn it all—all of it, all at once, usually completing the task by mid-April. Then they double-stock with cattle, largely imported, and most of that herd purchased with borrowed money. The fire-catalyzed prairies rapidly transform black char into green fodder. The freshening grasslands become an open landscape feedlot. By mid-July the fattened cattle are shipped to traditional corn-stocked lots before being dispatched to slaughterhouses. Relieved of intensive cropping, the grasses spring back and grow sufficiently to support another round of burning the following year.

The practice emerged out of 19th-century cattle drives, in which landowners burned at prescribed dates, under contract, so that approaching herds had pasture when they arrived. That early burning also prevented wildfires. Revealingly Oklahoma is among only two states that define fire legally within the concept of strict liability; there is no standard of negligence—if a fire escapes, regardless of reason, its setter is liable. (It's a code that permits easy access to fire, without the bother of permits and approved certification, and it works when embedded within a social matrix of burning.) What happens today is an updated version. The oddity is that open burning has persisted where, in most places, it has yielded to the enticements of industrial combustion. Fire has stayed on the land. A fire culture has endured.

That is the good news. The bad news is that the simplifying logic of industrial capital has applied its typical reductionism and transmuted fire from an agent of historical diversity into one of contemporary homogeneity. The system succeeds in bolstering the production of Black Angus and Hereford, but it works against the proliferation of the indigenous forbs, horned lark, and lesser prairie chicken. Tallgrass prairie, too, has its temporal patterns. It sprouts, ages, promotes, and stifles, yearly altering its structure and composition. It has its pioneer species and its old growth. Across landscapes it displays a mosaic of patches. The cornucopia of patches encourages a proliferation of niches and niche-specialist mammals, birds, and insects.

So, too, modern science has tended to parallel the logic of modern production, and it often views the economy of nature as it does the economics of commodities. Range science has isolated and studied precisely those critical components that have boosted the conversion of prairie grass into saleable meat. It scrutinizes each part of the prairie separately—that's what putatively grants it status as positive knowledge. It knows something of what grazing does; it knows something of how fire behaves; but until recently it has not sought to put grazing and fire together organically, which has left its Enlightenment-derived epistemology ignorant of what hominids on grasslands have known since the days of *Australopithecus*. The two processes don't act separately: they act together.

America's largest patch of protected prairie resides where the Flint Hills poke southward into Osage County. In 1989 the Nature Conservancy purchased the 29,000-acre Barnard Ranch, converted it into the Tallgrass Prairie Preserve, gradually added another 10,100 acres, and began the tricky process of regenerating something like presettlement prairie out of grazing-sated ranchland. They burned, they tore down fences, and they introduced bison. The preserve's bison could now free-range over some 24,000 acres. Its fires, however, were herded into a patchwork of biotic corrals.

Managers recognized that saturation spring burning might work for cattle, but it would wipe out everything that needed cover, nest sites, and the prairie equivalent of old growth. So they burn on a highly variable rhythm, set by the availability of fuels, which in turn results from the three-body interplay between bison, burning, and grass—a "messy" landscape, as the preserve's manager, Bob Hamilton, puts it. Moreover, the burns vary by season. Some 40 percent of the land burns in the spring, another 40 percent in the fall, and the remaining 20 percent in the summer. The land becomes a palimpsest of patches. There are patches for prairie chickens, patches for grasshopper sparrow, patches for invertebrates, and patches for bison. The patches are not fixed—there are no inscribed blocks fired with metronomic rigor. The burns occur when fuel is adequate.

There are many features of the landscape that attract and repel bison, and without fences to hold them, the bison have a lot of choice where to feed. But fire trumps them all.

Overwhelmingly, as hunters have known for eons, grazers go to the fresh fodder springing up after burns. Greening tallgrass gets cropped as soon as it surfaces, leaving the appearance of a mown lawn. So rich in protein is the grass that the older stems, yellow and waving in the wind nearby, are ignored. Nor must managers put out supplementary feeds to help the stock survive winter. The medley of burned patches keeps fresh fodder on the land year-round; and bison gorge on the low forbs—traditionally dismissed as weeds—that flourish amid the mix of grasses. By mostly avoiding last year's burns, bison allow the tough grass to remain, and by the third year—its old growth phase—it is actively shunned. These are the fuels for the next round of fires. The outcome is a rhythm of fallowing, a kind of swidden, intermediated by free-ranging fauna. So, too, the other creatures—the ecological specialists—can thrive. Greater prairie chickens can no more live in old-growth prairie than grizzly bears can in old-growth forest.

This is how the basics had evolved over millions of years. What aboriginal Americans did was to expand vastly the range of this dynamic and then to hold it against climatic pressures that sought to contract it. Their burning defined the ecoregion. What ranchers then did was what farmers have done with wheat and foresters with loblolly pine: they simplified, homogenized, and maximized for a single purpose. What the newcomers at places like Tallgrass Prairie Preserve are trying to do is to reverse, or more properly modernize, that process so that the land can recover, retain its historic character, and restore its biodiversity. In time, it is possible that ranchers may emulate the researchers and introduce patch-burning into their commercial landscapes. Experiments suggest that, amid those patches, ecology and economics may find a common cause.

What they all share is a recognition that prairie must burn. Where they differ is their preferences for what creatures will live on that burned land. What matters on this score is not whether the prairie burns, but how. The seasonal smoke that billows upward signals the character of that chosen regime.

OIL PATCH

At Osage County, however, the fire regime also extends downward. The sandstone strata that, warping upward to the surface, underwrites the Cross Timbers soils serves as below-surface reservoirs for petroleum. For a while, during the 1920s and 1930s, the region was America's primary producer of oil. Pump jacks still dot the landscape like the acacias of an industrial savanna. The bison herds on Tallgrass Prairie Preserve wander past them as they would scattered copses.

Few places display so dramatic a link between the two grand realms of combustion that define the modern Earth. The free-burning flames that recycle prairie must operate within a larger matrix of closed combustion that defines how Americans live on their land. It makes sense to reclassify the bioregion as a pyroregion, with living hydrocarbons on the surface and lithic ones below. Here, primordial geologic landscapes are exhumed, brought to the surface, and burned. People travel over the preserve on roads bulldozed and graded by internal-combustion machines. A parallel network of pipes carries fuel to the pumps and transports oil and gas from them to storage tanks, while power lines span the horizon. The spring cattle drives arrive by diesel truck and locomotive. Residents—managers, tourists, researchers—seasonally trek to the site by automobile and rely on natural gas and electricity from oil-fueled power plants to provide the energy they need to run the place. They even burn with diesel-and-gas-fueled driptorches dragged from ATV quads.

Contradiction or paradox—the reality is that the preservation of wild lands is something industrial societies do, and they reinstate them by the pyrotechnologies that prevail at the time. As roads and plowed fields broke up the range for free-burning fire during settlement times, so the grids of industrial combustion are allowing for their reinstatement, the regeneration of a patch-burned landscape. The resulting scene is a hybrid.

History, too, has its patch burns.

Those pump jacks and power lines get airbrushed out of most images. What visitors and donors want is a simulacrum of the lost prairie, now recovered. What is less appreciated, and less forgivable, is the way that people get airbrushed out of fire science. If fire ecology failed to bond flame and bison, it has failed far more miserably in bonding the animating flames to people. Yet people carried the torch that made the complex work. They still do. Some are managers; some are researchers; some, fire's accidental tourists littering roadsides with embers.

The reasons for eliding humans out of the prehistoric prairie is understandable within a national creation story that speaks of a wilderness America colonized by a civilized Europe; it is myth, and myth sings its own truth. It is less understandable for a science of ecology that purports to explain a natural world for which myth is not a prime mover. Yet until recently that is exactly what fire ecology has done: it has systematically stripped fire-powered biotas of their keystone species, the sole creature who has held a species monopoly over fire. Nature didn't burn those historic patches. People did.

Imagine a prairie in which bison graze preferentially on burned patches—and could themselves kindle the patches they wanted. Could anyone seriously erase that practice from a description? At least with prairies, now shrunk to a nanoniche of their historic dimensions, the case for restoring fire is clear, and there is really no option other than for people to do it. But its significance is ignored. It's as though people do it now much as scientists do it on experimental plots, as a surrogate for what might happen "naturally." The presumption is that the ideal system could evolve ultimately so that people would vanish from it.

In places—the Tallgrass Prairie Preserve among them—the role of humans as the Earth's signature fire creatures is accepted, or at least finessed. The land needs fire in particular patterns; managers do it.

Elsewhere, the argument is trickier, and acceptance comes only spasmodically, particularly where some aboriginal or pre-European presence is mandatory for political reasons. Nowhere, however, has fire science sought to grapple with the link between humanity kindling fires on quasi-natural landscapes and humanity rerouting its firepower through industrial combustion. Yet they overlie each other historically as fully as the burning prairies and the subterranean landscapes of Osage County do geographically. What is missing is the dynamic link between them.

What is missing is people. People set the overwhelming majority of fires in the past, and they set them today. The further east the prairie patch extends, the more it depends on humans to do the kindling. And that is no less true historically: the further back, the more prominent the role of human firebrands. Prairie managers have come to grips with this fact, pragmatically, if not philosophically.

Fire science still struggles. It's hard to claim objectivity when the author is also the subject. Humanity makes an unreliable narrator. But to ignore the keystone species for an ecosystem would be untenable for any other topic. An interest in traditional ecological knowledge is wedging open some intellectual space to explain the distant past. It's the present and future that remains inchoate. Here the wedge is the concept of the Anthropocene, which has confirmed the global, geological reach of humans. Ignoring people is like brooding over climate change without considering the human contribution to that change. What underwrites humanity's power, however, is fire. The planet's keystone species for fire—its fire monopolist—turned from burning living landscapes to burning lithic ones.

Combusting fossil fuels is the greatest transition in planetary fire history since *Homo* seized the planetary firestick from lightning. The oil patch testifies to a radical reformation of earthly fire. The switch to fossil fuel combustion as a source of firepower did not merely add another combustion realm; it began competing with the others. By technological substitution and outright suppression, typically by the instruments of its own contrivance, it has swept anthropogenic fire from the landscapes of most industrialized nations, and, for a long time, it sought to extinguish all naturally ignited fires even in nature reserves. Later, when fire's catalytic role became apparent in such places, advanced nations have yielded more room to lightning fire, even as they have intensified their determination to prevent and swat out human ignitions. Yet this extraordinary pyric transition is nowhere in the dominion of fire science. There is no perceptual link between the two realms of combustion, a failure of imagination so massive it casts the entire enterprise of fire ecology into question.

At Tallgrass Prairie Preserve, however, the two worlds coexist in the same field of vision. Pump jack and bison, grass and oil, flame whipping in the wind like whitecaps and diesel-fueled pistons pounding without pause—they all converge, and they demand a full-spectrum appreciation of fire's contemporary ecology. Those pump jacks are the outcrops of industrial fire's deep ecology. Humanity's shift in fire practices has indeed unbalanced climate, but it has also destabilized whole biotas, now being rapidly remade by the means and to the purposes of modernity, and it has further unhinged landscapes by altering the fire regimes with which those ecosystems had long come into accommodation. One common presence stands behind all these manifestations—humanity as a fire agent.

Tallgrass Prairie Preserve runs on both combustions. It requires gasoline as much as fallowed grass, and it accepts both as a practical necessity. Fire science has yet to catch on—or to catch up.

WOODY PATCH

The road to understanding is there. In fact, there are roads all over the place, many paved with asphalt from the oil patch and all populated with a mechanical menagerie of petroleum-respiring machines more far-ranging than the prairie chicken and more prone to mass into herds than bison. Those roads converge ultimately on cities, and together roads and machines constrain the prairie and its flame.

They chop up the indigenous geography into awkward new patches; they alter the rhythms that long characterized the seasonal migration of prairie denizens; they replace anthropogenic burning with internal combustion and its byproducts. Against prairie they both push and pull. They push by constraining the breadth and capacity of prairie to carry free-burning flame. By roading and plowing, they break up the continuities that, paradoxically, had allowed for the ceaseless churn of patch-burning and grazing. They pull by powering two woody sprawls that are taking over more and more of the extant grasslands. One is a spread of trees, particularly exotics like the eastern red cedar; the other is an expansion of wood-framed, or at least wood-stuffed, buildings. Each, in its way, is replacing grass with woods. Both are the expressions of a civilization propelled by industrial combustion.

Like other invasives, eastern red cedar (*Juniperus virginiana*) perturbs the fire regimes of the landscapes it occupies, and like the others it can seize those sites because they have already been perturbed. It is, to the Great Plains, what cheatgrass is to the Great Basin. But unlike cheatgrass, which propagates amid promiscuous burning, red cedar roots where fire has lapsed. Once established, it is difficult to ignite, but when it does catch, it burns with a longer and more sustained ferocity, the prairie equivalent of a crown fire.

In its seedling stage it sits in grassy thickets like a plum in a pudding: a fire of even average intensity can easily strip and kill it. A flaming front can engulf the entire tree. Whether such fires sweep a field depends on how closely cropped the grasses are and whether or not a fire gets set. As it grows, cedar lifts more of its canopy above the grass, and, equally to the point, the transient flames from the close-grazed understory grasses no longer hold enough fuel to kindle the more stubborn needles. The lower skirt of the cedar may burn or scorch, but the bulk remains, now partly immunized against another surface fire. The infection spreads exponentially—a spot here and there of cedar, then a dappled landscape speckled with young trees, and suddenly a forest. At this stage there is insufficient grass to carry a continuous front of fire. The landscape has flipped into an alternative stable state, like cheatgrass in Nevada, alang-alang in Sumatra, or linden in central Europe.

Only the most intense fire, or the combustion of fossil fuels for grubbing or spraying, can dislodge those new woods. That requires ample kindling in the form of grass, perhaps after a season of unusual rains and lessened grazing. It requires winds like a cyclone to allow flames to lengthen and leap from one blazing tree to another. And it demands a precisely timed ignition. In the realm of industrial landscapes, that combination of cards is as rare as a full house, and as residents cluster into cities and remake the rural countryside into exurbs, they perversely welcome cedar as a windbreak, as a shielding screen for privacy, and simply as a tree on the plains, which holds its own totemic values.

Deliberately and indifferently, the eastern tallgrass prairies are becoming thickets of eastern red cedar. Grass fires become less frequent, and when they come, they are mixed and savage—a prairie fire on woody steroids.

Then there is the city. Here lies the social complement to the laissez-faire conversion of grass to semiferal woods: the deliberate construction of wood-framed houses or wood-stuffed residences. The two processes are linked, as city and cedar merge into a collective complex of wood. It is a composite that eerily echoes the sinuous geographic scrawl of the Cross Timbers with its contours sculpted by the peculiar edge ecology of internal combustion. Cedar and city are resequestering the carbon released by the oil patch into a massive woody patch.

The town as suburban sprawl is as much a product of an industrial economy of fire as the exotic cedar with which it congregates. It has laid down a new matrix for burning in which lines of fire track corridors of travel and fields of fire, such places of concentrated burning as power plants and residences burning heating oil. Vestiges of open burning endure; Oklahoma has in general tolerated private firing, though it is unforgiving of failure. But more and more, an industrial regime has absorbed and confined open flame. For those who grow up in a city, burning houses are something that happens on television, and open burning belongs beyond the urban fringe.

Or did. The areas bounded by a Greater Cross Timbers region have, over the past decade, added fire-destroyed houses to tornado-destroyed ones. The scene is bizarre, as though the tallgrass prairie had never been settled, or was being unsettled, and Oklahoma had morphed into California, and Midwest City and Choctaw had begun to channel Malibu and

Lake Arrowhead. People behaved not as they historically had, when they had survived and tamed prairie fires, but as they saw on TV screens today; fleeing, standing stupidly on combustible roofs with garden hoses, texting and jabbering into iPhones. They turned, as they must, to the implements of industrial fire, from autos to fire engines, in order to flee or fight.

A specter that society had thought long banished, maybe extinct, returned. It was like watching the revival of pandemic tuberculosis or the emergence of a drug-resistant staphylococcus, as though a scourge from fetid Third-World tropics had established itself. There is scant reason for modern housing and suburbs to burn: the built landscape has become less combustible and can yet become more so. We know how to keep roofs from burning, how to protect exterior walls from heat and flame, how to design yards that shield against fire rather than propagate it, how to protect people. Last-minute flights in cars over crowded, smoke-obscured roads don't do that. Nor does erecting windbreaks of red cedar, ready to saturate a downwind house with ember showers. Nor does permitting combustible roofing. Nor does allowing one hazard to sit next to another, so that fire can jump from one to another without regard to landscaping in between. We know all this—know how to encode such knowledge into law and custom—yet have allowed the woody patch to sprout and tenaciously propagate without taking remedial measures.

Instead, the borders of the two fire realms have slammed together like the earthquake-prone frontier of two tectonic plates, ready to erupt with an occasional Richter-scale fire. The patch burns of history have somehow burned through the overlays of contemporary life.

WHEN THE WIND COMES SWEEPING DOWN THE PLAINS

For days, all over eastern Oklahoma, on commercial pastures, public lands, and private preserves, fires had daily sprung up, sprinted over hills and swales, and sunk into oblivion with the dews of evening. It had been a mild, wet, windy spring, and pastoral burning was slightly behind its normal schedule. On April 8, 2009, taking advantage of favorable conditions, the burning had continued into the evening. The next day, Friday, had a cold front forecast, and most burners stayed their hand. But some did not, and there are always stray sparks. Like black rats, houseflies, and litter, flames travel with people.

A few late-morning fires became many. The gusting winds fanned errant spot fires into fire fronts and whitecaps of flame into a tsunami. The normal logic of fire spread and firebreaks collapsed as swarms of sparks flew locust-like over normal barriers and feasted on fresh fuels of dry grass, field-encrusted cedar, and vulnerable houses, and smoke columns bubbled up like thunderheads. By evening, April 9, fires had rampaged over a swath of central Oklahoma. An estimated 100 homes had burned; several thousand residents had been evacuated; one firefighter had died. Governor Brad Henry declared a state of emergency over 31 counties.

Along that exurban fringe, and in some cases well within it, two visions of landscapes collided. Each had its own relationship to open fire.

The city wanted no open fire and eagerly traded grassy kindling for woody fuel. Urban fire services would gladly banish fire. The city's response to the outbreak was for residents to flee and for firefighters, outfitted with the apparatus of industrial combustion in the form of engines, pumps, and aircraft, to attack the assaulting flames. Given the extraordinary pace of the burning and its capacity to fling firebrands hundreds of feet ahead, this was a quixotic quest. It was not possible for a fire militia to scale up as quickly as the flames had, and there was no way for it to contain a fast sprawl as extensive as the slow sprawl of urbanization. The firefight was telegenic, instantly transported into a virtual world that became digitally visible and engaged even those far removed from its searing heat and obscuring smoke.

By contrast the countryside demanded fire but wanted it on its terms, not that of arsonists or accidental incendiarists. One of the breakout burns threatened the southwestern borders of the Tallgrass Prairie Preserve. It was not the fire in and of itself that mattered—the land would be burned anyway. What mattered was its regime; the preserve had a rhythm of patch-burning and didn't want it broken if possible. Quickly the staff at the preserve responded, much as their city counterparts did, with diesel-fueled vehicles, quads, and pumps. But they did not attack the flames directly. Instead they picked out a graveled road far in advance of the front, laid down several passes of water on the upwind side, and then drove quads along the downwind side and stripped out lines of fire with driptorches. There were no overhead TV action news cameras, no panicked residents scrambling into the nominal safety of SUVs and the open highway, no loose similes by which to remake a quasi-natural event into a semblance of a terrorist attack. They met feral fire with tamed fire.

The city saw fire as an intrinsic threat, responded to the crisis as a social disaster, and dispatched an all-hazard emergency service. The country responded to the threat as a problem in land management. A wildfire was a familiar problem to deal with, like a hailstorm, not a specter from an alien world. The city media reported on and analyzed the fires with the same breathless and hackneyed language used for the fires outside San Diego or the Angeles National Forest. Drought. Fuel buildups. Wind. These were the putative causes, though fires traditionally burned at this time of year, which was a period of plant dormancy, and fuel and drought had meaning only when applied to eastern red cedar, not a year or two's growth of grass. So, too, the responses were clichéd. Drop water from aircraft. Send engines. Evacuate. The public mood was shock, mingled with a sense of voyeuristic self-regard—shock that such fires had struck Oklahoma and a macabre touch of vanity that Oklahoma could claim fires that normally cavorted on the California coast. The countryside saw those flames as part of living on the land. It suffered no burned houses, reported no lives threatened, endured no cindered assets. Fire control was the flip side to fire use. Both were matters of tending land, of a piece with watering fields, cleaning ditches, or cropping surplus bison.

The difference matters because democratic politics derives from its citizenry, and even in Oklahoma most citizens reside in cities and suburbs. More and more of the state's inhabitants know fire only as refracted through TV, radio, podcasts, the Internet, tweets. They do not heat their houses, cook their meals, read and tell stories over a fire. They use flame less and less on their yards and find it difficult to imagine its semidomesticated use in remote pastures and nature preserves. Yet urbanites will determine the ultimate fate of fire on the land. The fires they witnessed were the ones that leaped I-35 and Highway 51, that rushed over empty lots and unkempt pastures on the city fringe, that sent smoke like a squall line through city centers. The fires seemed, as one begrimed fireman in Wellston exclaimed, like "hell on Earth." His counterparts on the preserve might instead imagine the nominal conflagration as a poorly lit patch burn that had temporarily slipped its leash.

Then they ended. However fierce, grass fires, even those fed by woody supplements, burn fleetingly. They pass with the wind. Within hours the flickering front had blown itself out.

That is not likely to happen, however, with the contemporary fire regime. It rides the historic front between two realms of combustion, and until its gusty passage ends, the fires will continue to move from grass to woods. Whether postwar houses prove as durable as the post oaks and blackjack oaks of the Cross Timbers remains to be seen.

FIGURE 1. Prairie researched: Konza. Part of the matrix of plots, each with a different burning regime, near the preserve headquarters.

FIGURE 2. Prairie restored: Tallgrass Prairie Preserve, with free-ranging bison. Fauna are critical, but few sites have the scale required to accommodate a full palette of browsers and grazers.

FIGURE 3. Prairie retained: Arrowwood National Wildlife Refuge amid the prairie potholes of North Dakota. Upland and wetland—the geographic equivalent of the wet-dry rhythms that climatically underwrite fire.

FIGURE 4. Prairie rebuilt: Nachusa Grasslands. Something akin to the indigenous prairie constructed painstakingly out of cornfields, catalyzed by burning.

FIGURE 5. Domesticated trees: the Nebraska National Forest, planted, then held against fire from the Sandhills grasses. Courtesy U.S. Forest Service.

FIGURE 6. Feral trees: mesquite, the iconic scrub of the Texas savanna. Fire in grass can prevent its establishment, and fire can hold it in check, but once established fire alone cannot drive it off a site.

FIGURE 7. (a) Henry Graves, future chief forester, watching light-burning at the Black Hills, 1898. Source: U.S. Forest Service. (b) A century later, light broadcast burns yield to intense pile burns. The Black Hills pine grows faster than it can be culled by axe and fire, and both pile and broadcast burning lag.

FIGURE 8. Brush held at bay: Wichita Mountains, a mosaic of grass, scrub, and granite.

FIGURE 9. Brush unchained: eastern red cedar, eastern Oklahoma.

FIGURE 10. Pleistocene, Pyrocene: (a) aerial of prairie pothole region in North Dakota; (b) fracking on Bakken shale, on a scale easily viewed from space and an eerie echo of patch burning amid the potholes. Source: USGS; NASA.

FIGURE 11. Old fire, new fire: (a) Fighting a prairie fire in Kansas, 1874. Source: *Harper's Weekly*. (b) Pump jacks frame the scene at Tallgrass Prairie Preserve. So, too, does our industrial order sketch and define the borders and dynamics of grass and bison.

TEXAS TAKES ON FIRE

A Reconnaissance

PROLOGUE: FIRE AND TEXAS

I knew that sooner or later I would have to have a go at Texas, and I dreaded it.

—JOHN STEINBECK, *TRAVELS WITH CHARLEY IN SEARCH OF AMERICA* (1962)

In this world nothing can be said to be certain, except death and Texas.

—WITH APOLOGIES TO BEN FRANKLIN

TO DATE, ATTEMPTS to write the national history of fire in America have managed to ignore Texas. What is particularly odd is that no one noticed. For all its geographic and demographic heft, its economic clout, its intrusive politics, its boisterous self-promotion, Texas simply didn't register. Fire in Texas raised no more notice on the nation's consciousness than a bevy of stray goats or a patch of mesquite. For that matter, the topic had barely registered in Texas. Fires there were—and always had been; the first fire reported by Europeans in North America involved fire hunting for rabbits by Texas Indians around 1529. Thereafter, as recorded, fire seemed to influence Texas little, and Texas did almost nothing to shape national programs or policies.[1]

The void is all the more striking when compared to Texas's determination to announce itself to the nation so grandly in most matters. By its sheer size, its southern pineries, and its western grasslands, it had abundant burning. The first national text on forest fires was written by A. D. Folweiler, who subsequently served for 18 years as director of the Texas Forest Service. The first survey of fire ecology in North America was coauthored by Henry Wright, a professor at Texas Tech University in the early 1980s. Yet neither text moved the Texas fire scene into national prominence. There were no mechanisms to apply their examples and ideas in ways that could make the state an exemplar. The absence of public lands severed a natural conduit for nationalizing events. The state boasted no celebrity landscapes to advertise its practices. What Texas boasted of was Texas.[2]

The new millennium upended this comfortable narrative. A new Texas had morphed from the old, and fires ripped through it like monstrous tornadoes. Texas could not contain them and turned to the nation for assistance. Whether sought or not, what had long remained an intrastate soliloquy has became a national conversation. Texas has had to take on fire. Which means it's time for fire history to take on Texas.

LONE STAR BURNS BRIGHT

In 1988, a year later made famous by extended conflagrations at Yellowstone National Park, two March fires in Callahan County, Texas, swept over 336,000 acres and burned winter forage, some cattle, and miscellaneous field structures. In 1996, however, a 20,000-acre fire northwest of Fort Worth consumed 65 houses and 90 outbuildings and left 52 firefighters and civilians with injuries. Something had changed.

Fires in Texas were normal, as abundant as grass. In fact, though it was not apparent at the time, the 1996 season inaugurated a cavalcade of serial conflagrations in which each startling exception established a new norm. Year by year—1998, 1999, 2000, 2001—the fires flared with unsettling intensities, swelling in ferocity and damages. No longer were they content to burn grass and brush: they began to crash into small towns and incinerate houses. Then wet conditions returned, which caused the fires to abate but only bolstered the stockpiles of combustibles that powered them. On December 27, 2005, wildfires returned and swept into the Cross Plains, where they consumed 85 built homes, 25 mobile ones, 6 hotel units, and the First Methodist Church; two people died. The drought strengthened, and as it squeezed the land, fires squirted out from its grip. In March wildfires ripped across the Panhandle, with the two largest complexes together burning over 900,000 acres. I-40 was shut down from smoke, eight towns were evacuated, and 12 people perished. In 2008 fires crashed through the Hill Country—the fabled Heart of Texas. On April 9, 2009, some 205 fires, driven by 70-mile-per-hour winds, blew up across 14 counties in north Texas; the most damaging ripped gashes through Montague County where they destroyed 86 homes and killed 4 people. As previously, high winds grounded any kind of air attack. Earlier, on February 28, 2009, a relatively small burn (1,491 acres) in Bastrop County consumed 26 homes, 20 businesses, and prime habitat for the endangered Houston toad. The Wilderness Ridge fire was so sudden and savage, however, that it prompted a worst-case study from the Texas Forest Service under the assumption that the outbreak represented a new maximum. It didn't.[3]

On November 15, 2010, the first of a rolling thunder of wildfires commenced that crashed over almost every region of the state—some 29,394 fires that burned 3,933,436 acres, almost half of all the land burned by wildfire in the United States. Unquenchable fires broke out east and west, north and south; through Possum Kingdom, across the Panhandle outside Amarillo, between Longview and Mount Pleasant in the northeast, over the Davis Mountains beyond the Pecos, and in a thick swathe of scars along the 100th meridian. But Bastrop County, east of Austin, within eyesight of the state capitol, absorbed the worst, as a virtual red norther, powered by gusting dry winds from land-falling Tropical Storm Lee, drove fire as straight as a gunshot across 34,356 acres and took 1,645 homes and 2 lives. Over the course of a decade what had been considered Texas's worst fire outbreaks had effectively doubled with something like the trajectory of Moore's law.

For most of the 20th century, while the United States erected a fire protection system that bonded local and federal agencies and ambitions, Texas stood apart. Lacking federal lands, it did not integrate into the evolving national establishment for either operations or research. It had its own way, adapted to its peculiar circumstances, culture, and sense of itself. Moreover, while fires were common, bad fires were rare. Texas, all of Texas, had always burned, but by the time a national infrastructure emerged, Texas had beaten back—literally eaten back by grazing—its pioneering fires, had reduced fire management to firefighting, and downloaded responsibility to local jurisdictions. What fires continued never figured in the national census. What happened in Texas stayed in Texas. Within the national dominion Texas had created in effect a separate fire fiefdom. Texas took care of its own.

But as its fires claimed more and more of the nation's attention, and as they insistently exceeded the state's capacities, Texas found itself inexorably, if reluctantly, drawn into the national grid. It could no longer cope by itself. The 2011 drought and fire season had struck with the power of a category 5 hurricane. The Texas Forest Service estimated that together they had taken out perhaps 2 to 10 percent of the state's trees. Every county of Texas's 254, save one, had been declared a federal disaster area. Of the 13 largest fires in the United States, Texas claimed 6. Neither Texas nor the United States could afford to ignore the other.

The 2011 fires were a subcontinental calamity that went well beyond Texas. Wind and drought spawned fires no one had witnessed before. No fire agency alone could have coped alone with such an outbreak. Record fires blasted across federal lands in Arizona and New Mexico under a U.S. Forest Service that prided itself on being the best fire organization in the world. But Texas did not have a controlling agency for fire or a systemic program, only a brokered set of arrangements by which to respond to emergencies. This time Texas could neither shield itself from the national scene nor contain the fires within its own borders. Amid the constellation of big burns that inscribed itself across the Greater Southwest, the Lone Star state shown like a supernova in eruption.[4]

All parties learned that while Texas is big, fire is bigger.

WHERE THE SOUTH MEETS THE WEST: THE TEXAS PERSUASION

Texans have always been quick to tell others that its people define Texas, not its geography. Its sense of identity resembles Brazil's, a country whose bigness is an important part of its consciousness but whose society is defined by its social relationships and stories about its origins.[5]

Certainly in its pyrogeography there is little distinctive about Texas. While the state's landed estate is diverse, its vaunted contrasts mostly involve small climatic gradients made large by distance. The state is coastal and humid in the southeast, and dry and slightly higher in the northwest; its far west fragment belongs with the Southwest's sky islands. What has determined its physical geography is its political history, which set its borders. What matters for fire is how Texans have chosen to live in that setting.

Texas is where the South meets the West. The state is anchored, historically and culturally, in its humid east, where prairie and piney woods continue a southern landscape and where American settlement entered and took root before spreading west. The story of how that happened hinges on events of 1835 to 1836, when Texas empresarios and freebooters found common cause in rebelling against an enfeebled Mexico. The revolution led to the Republic of Texas, which lasted 10 truculent years, fighting against and launching incursions against Mexico and Plains Indians, until a de facto American protectorate led to admission into the Union, an act that provoked America's war with Mexico. A scant 13 years after joining, Texas again seceded by joining the Confederacy. By 1865, fewer than 30 years after it acquired an identity beyond being a tenuous province of Mexico, it again found itself among the United States. Those tumultuous events, particularly its first revolt, gave Texas a peculiar creation story.[6]

Yet other states, even large and powerful ones, have enduring origin myths. What translated the Texas story into political power was its fleeting experience as an autonomous republic, which granted it leverage to retain its public domain upon entering the Union. Only two other states kept their unsold lands; both (Maine and West Virginia) were carved from founding states under peculiar political circumstances. Even potential Indian reservations were moved north of the Red River into Oklahoma. Because its lands were extensive, Texas had to do what elsewhere was done by the state and federal government together, or for the far western states, mostly by the feds. Texas had an internal source of wealth and power uniquely its own.

Still its imperial reach was one the sparsely settled polity could hardly hold, much less grasp. Its institutions lagged hugely behind the scale of its land and the tasks that scene demanded. In 1836 it had a population of 212,592 and a landed estate that also included portions of what today are New Mexico, Colorado, Oklahoma, Kansas, and Wyoming. (Its present, reduced shape resulted from the Compromise of 1850.) The nature of the country, moreover, favored large-range ranchers rather than freehold homesteaders; and when settlement did thicken, it did so in dispersed forms that created no commons and scant collective interest. Other western states could survive because the federal government oversaw the unclaimed public estate and then kept much of it as permanent public lands. Texas had to find ways to manage on its own, which meant foregoing institutions in favor of big landowners, and later just its "big rich." (Interestingly, both West Virginia and Maine also fell under the sway of big landholders; mineowners for the one, logging companies for the other.) Texas institutions never matured along the lines they did nearly everywhere else. While the state certainly integrated into the country economically and became exceptionally adept at extracting concessions from the nation overall, its landed estate, which has granted power to Texas exceptionalism, did not follow national norms.

Instead, Texas had to assert claims rather than stake them. It reached where it couldn't grasp. It boasted where it couldn't control. It turned to big men, Texas Rangers, and quasi-legal vigilantes to assert order. It nurtured a popular culture exceptionally rich in folklore and folk inventiveness in place of an imported civilization. Its elites were premised on access to land-based resources. The weak institutions that characterized its founding became permanent features and were then branded into a sense of itself. Southern traits, particularly skepticism of authority and appeal to popular violence, found themselves no longer fenced in by even nominal agencies or social conventions. Where the piney woods thinned into the grasslands at midstate, even the constraints of agrarian ecology broke down. The South met the West and rode away untrammeled. The land was bigger than its inhabitants. An appeal to strong character and guns replaced the overstretched sinews of civilization. In some respects, the stampede unleashed by those early years has never been roped back into the national herd.

Much that is attractive about what might be called the Texas persuasion follows from those historical circumstances. Its openness, its tendency to personalize abstractions, its fresh and frequently figurative language, its often larger-than-life presence, its disdain for the enshrined and the solemn and its embrace of the entrepreneurial, its suspicion of self-serving elites in favor of individuals and self-reliance—these are laudable and likeable traits. But those same attributes can turn sour.

The preference for the personal can lead to a politics of personalities rather than of laws and institutions. The fresh may segue into the coarse. Self-empowerment and violence can morph into intimidation and redneck bullying. Suspicion of power can yield to conspiratorial brooding. A fixation with the big can become bombastic and boring.

A distinctive culture often means a culture that cannot move easily outside its borders. In speech, dress, and self-identification, the Texan is often a Texan wherever he or she is. Self-awareness can yield to narcissism and caricature. The Texas brand—the state silhouette—is everywhere. Everything is "Texas" this or "Lone Star" that. There are waffle irons and garden stepping stones in the shape of the Great State of Texas. No other state makes its profile into a marketing brand or has its expats referred to as "Utah" or "Wisconsin."

It takes little imagination to see in its past the reasons why country-western music thrives here, why Texas claims half the executions in the United States, why Texas politics fares badly beyond the pale of the Lone Star, or why fire in Texas looks as it does.

Texas evolved differently. Lots of states did, but few others were also large enough to throw their weight around or resist the homogenizing influence of the national culture. Texas was.

The Bear Flag Revolt in California never got beyond a homemade flag and a big drunk before the state was annexed to the United States and a gold rush established its formative character; the state imagined itself as wealthy from its origins. The Texas revolution, complete with a shrine, the Alamo, led to a separate nation. If its lands were too vast to consolidate, its bold claims sent down the taproots of an anti-institutional culture. When Spindletop opened the spigots, the mineral wealth flowed through familiar channels. Half of California was public land. All of Texas was state land rapidly dispersed into private ownership. California had extensive national parks and forests; Texas had giant ranches. California became integral to America's wildland fire establishment. Texas stood apart. It had to do at a state level what other western states did through federal institutions, which meant it did less. Because of its mixed land base, California could shape, and continues to influence, national fire policy. Texas could not, and it contributes toward national practice largely by its appeal for emergency assistance.

Preference, necessity, and historical accident made anti-institutionalism a Texas persuasion that put but a small premium on public goods and services beyond the most elemental kinds of protection, and even less attention where that public good pertains to environmental matters. There never emerged, from its immense estate, a wilderness movement or a passion for deep ecology or for an environmental concern beyond "beautification" since all those concepts smacked of interference with private property. While purchases have created a national forest in east Texas, a handful of national parks in the far west and along the coast, and a few wildlife refuges, a scant 2 percent of Texas is public land administered on behalf of the American people as a whole, a proportion comparable to that of Iowa. Texas thus avoided unwonted federal influence; Texas could tend its own ranch. It also missed integration into national fire institutions that would have allowed the state to share in the federal largesse for management and research and to mold the country's agenda.

That may change. Wildfire has joined hurricanes, earthquakes, floods, droughts, and tornadoes as a routine natural disaster. Texas is one of the three states most susceptible to such outbreaks (interestingly the other two are Florida and California, both centers of an identifiable fire culture; Texas may be joining them). Precisely because it is different, large enough to make those differences unmistakable, and subject to wildfire on a scale that can distort the space-time geometry of the national fire establishment, Texas can no longer be ignored or choose to isolate itself. Over the past few decades Texas money, often allied with Chicago minds, has loudly declaimed what America's political economy should be. Texans have not similarly addressed America's political economy of fire. The Texas experience in which responsibility is downloaded to the lowest authority is highly relevant to the renegotiation of rights and responsibilities that lies at the core of an emerging National Cohesive Strategy.

The best analogue to Texas is, in fact, not in the United States at all but in Canada, where provinces control their lands and natural resources and the provinces ally into a confederation rather than a national federation. Think Ontario, with its bulbous north, which proved as distorting to its ancestral Great Lakes heartland as an expansive west was to a Texas nurtured in the southern pines. No Canadian province, however, disdains authority the way Texas does. An acknowledgement of "good government" is enshrined in the British North America Act that congealed those colonies into a confederation; Canadians welcome institutions as a necessary feature of civilized life, and Canadians (like Australians) look to proper government to shield them from life's natural and social vicissitudes. At least in principle, Texans don't.

When wildfire slammed its boreal backcountry, Ontario cultivated a state-sponsored system of protection. When fires blistered its dryline west, Texas mustered posses, or, when conditions were truly awful, sent in the equivalent of a ranger company. The state could avoid the unpleasant politics of amalgamation because its fires were within the capabilities of its volunteer fire departments. Now, it seems, they are not. They may require the resources of the nation, not unlike the suppression of the Comanche and Apache in bygone days. That prospect may prove awkward for Texans, and until Texans reconcile and adopt those standards accepted by the rest of the United States, it may prove no less uncomfortable for the rest of the country.

THE TEXAS FOREST SERVICE RIDES WEST

Fire protection evolved, as so many other Texas mores did, by flowing east to west. When the Texas Forestry Association in 1914, and then the state legislature in 1915, responded to the Weeks Act, they followed the tradition of other southern states with pineries that were then feeding the ravenous mills of the American logging industry. Their collective focus was east Texas, its lower reaches dominated by the swampy Big Thicket and its upper by tough pines such as shortleaf. The Forestry Law authorized a state forester and a Board of Forestry at Texas A&M, and a Division of Forestry under the Texas Agricultural Experiment Station. The act even allowed for the purchase of land for nurseries and demonstration forests.[7]

A decade later the board had bought its first state forest, organized forest protection (which meant fire control), and in 1926 renamed itself the Texas Forest Service (TFS). By then oil had replaced lumber as the state's most valuable export commodity. Its woods were so badly mauled that the state authorized, in principle, the purchase of cutover lands for reforestation as state forests. A decade later the board established a forest products lab. In effect, the state was following the model of the U.S. Forest Service. It was a big operation for a state but much less than the federal program, particularly with the CCC remaking the national forests and parks. By 1934, with the timber economy slashed and the best lands scalped, the U.S. Forest Service acquired a patchwork quilt of lands for rehabilitation. In 1948 the TFS was granted agency autonomy from the university, although it remained affiliated. A year later it hired A. D. Folweiler, formerly of the USFS and the author of a text on forest fire control, as director. In 1950 the state experienced what, for a century, it regarded as its fires of record.

The TFS continued to mature. It was a small outfit. It understood its mission and colleagues. Its fire responsibilities were parceled out among paid and especially volunteer fire departments, stiffened like concrete with rebars by ties with the federal Forest Service. Overall, Texas remained an agricultural landscape of farms, ranches, and small towns. Among the piney woods it was densely enough settled that, along with logging companies and state and federal employees, fire protection was possible without the kind of expansive campaigns typical in the public lands of the West. Moreover, the land was broken; fires soon struck impermeable borders. Even big blazes rarely blasted beyond a single burning period. When the timber industry regenerated in the 1960s and 1970s the TFS grew with it. But it remained a cozy, if skilled, agency, tailored to regional needs and a state committed to small government, whose legislature only met on alternative years. Its hearth was east Texas. Even the Lost Pines of Bastrop County was a stretch.

To the west the lands spread out, the population thinned, and both pastures and fields relied on wells to hold them against aridity. With time the farms spread, and ranching became less open range and more like farming. The farther west, the more big ranches and natural range prevailed. The gradient of settlement followed closely the gradient of climate. This held for the United States overall, of course: the Texas tale is a cameo of the country. In fact Texas is unique among the plains states in that it stretches from the humid woods of the east to the arid grasslands of the west. Texas, however, spanned east and west without the national resources and institutions that could be brought to bear on, say, the West River country of South Dakota. The state's writ for fire protection barely reached the Balcones fault.

In his famous study of the Great Plains, by which he really meant Texas, Walter Prescott Webb pondered the encounter of a woods culture with the plains, by which he meant not only grasslands but aridity. It halted homesteading in its tracks. The shock forced a suite of adaptations in fencing, watering, fuels, and livestock husbandry. The southern plains even came with its own winds, as distinctive as Southern California's sundowners. The dryline traced the northward flow of air into the state. Where Texas fronted the Gulf, the air was moist, and where it bordered Mexico, the air was dry, and the rude frontier they shared was prone to gusts like miniature cold fronts that could readily boost stray sparks into sweeping fire fronts. So the need to adapt settlement to novel conditions for water and fencing proved no less true for fire protection.

The TFS had prudently stayed well east of the dryline. The nature of modern Texas settlement, however, has forced it to cross that invisible line in the sand.

Like many crises this one had crept along silently for decades before bursting into view. Most of the country had spent two decades fretting over exurban sprawl and megafires. With a few exceptions Texas had not noticed or had seen only a booming demographics and home-boosting economy. While specialists fretted over scenarios in which a conflagration might strike the Texas exurb as it had many western states, the prospect of fires actually rushing in like long-vanished Comanche raiders seemed to most residents a fantasy.

The fires of the old frontier, like its free-ranging prairies, had long ago faded into memory. Big burns, like dust storms, seemed a specter of the past. Throughout rural Texas fires were as abundant as mesquite thickets, but they were a nuisance, not a serious threat. They were the combustion equivalent of eaten-out pastures. In 1894 a grassfire, with little to stop it, burned across the Llano Estacado reportedly for "several million acres." The 1950 fires burned 158,319 acres within TFS jurisdiction but through many small burns; and while no doubt much more burned beyond its writ, the outcome was nothing like frontier times. The Texas grasses were grazed down and trampled; the prairie was shattered, much of it farmed, and all of it roaded. The great 1956 drought only lessened the available fuels by retarding grass growth (76,171 acres burned). What fires occurred were of local concern only. Even on giant trans-Pecos landholdings ranchers handled them as they did other threats and breakdowns. Almost nowhere did fire ramble long or far or intrude on the state's, much less the nation's, political consciousness.

With quickening tempo, however, a new frontier had emerged to reshape Texas. An economy based on industrial combustion was refashioning the landscape no less than its occupying society. Allowing for the climatic gradient—more ranching and larger holdings in the arid west—the old settlement style had imposed a remarkably similar pattern of land use by a population of common ethnicity and political identity. Gradually, as farms replaced more ranches, the land became even more inhospitable to fire. Flames could hardly begin before they hit a plowed field, a grazed paddock, or a graded road. Grassfires burned briskly but without much power: there simply wasn't enough fuel to loft them over those barriers. The landscape had been broken to plow and hoof, and then to automobile and asphalt.[8]

But as the human economy changed, so did the natural. The separate tiles of the grand mosaic were plucked out and replaced. The big story was, the human population bloomed, gathered into cities, and fragmented. In 1850 Texas had a population of 212,592, and in 2010, 25,145,561, a hundredfold increase. Demographic clusters became the norm as cattle moved into feedlots, and people into metropolises. By 2000 Texas claimed three of the country's top seven urban centers, and even modest towns were spreading outward like locoweed. Exurbs pushed into dried swards of grass, mesquite, oak, juniper, and pine; they were tethered to subterranean reservoirs of oil as an earlier settlement regime had been to water. Big holdings broke up as generations moved to town and law office, or held parcels as sentimental mementos or arriviste trophies. The small holdings could thrive only with outside income. Lightly grazed and unplowed lands ripened into grasses and brush, lay fallow under conservation easements, or morphed into amenities properties. Ranches raised deer or exotic fauna for hunting rather than only livestock for beef. That middle, broadly rural landscape that had informed Texas fractured, and a fraction passed out of intense production. By 2000, 55 percent of Texas remained ranchland, but active ranchers were aging (on average over 55), many ranches were really recreational ranchettes, and 70 percent of holdings had fewer than 50 cattle. While 80 percent of Texas remained rural, only 17 percent of its population was. In the West the public domain offered a counterweight to similar fragmentation; in Texas there was no comparable resisting force. The old Texas was going, and it was taking its once-fenced fires with it.[9]

The new Texas spawned a bolder, more exotic firescape. There was a less bounded landscape to burn, plenty of ignitions, and a settlement pattern that was hardly a wildland-urban interface since there was no genuine wildland and little urbanization, only a strengthening of feral countryside and of slouching townsites and the blurring of a porous border between them. While intensive grazing remained the norm, more parcels were free to boost fuels, and they could act like unvaccinated members in a herd, potential vectors for combustion contagion. Fuels revived as brush and fallow grass, while institutions struggled to keep up. Certainly this was true for fire protection arrangements ("system" is too strong a term). Outside of cities and the piney woods fire control was the purview of volunteer fire departments and county officials. It remained only to add to that open combustion chamber a climatic anomaly that could drain fuels of moisture, bring power lines down in windstorms, and hurl successions of cold fronts or performance-enhanced drylines across the southern plains to spark wildfires like a recurring nightmare.

Beginning in December 1995 and continuing for 515 days, this unstable frontier erupted in flames. With startling savagery, wildfire swept into the Cross Plains, a rural community southeast of Abilene, and burned 6,835 acres and 110 homes while the Canyon Creek fire took 47 homes on a mere 130 acres. The new year sent four other fires through north Texas for 84,000 acres and another 130 homes, some between Dallas-Fort Worth and Wichita Falls. In March 900,000 acres burned in the Panhandle and destroyed 89 homes. This seemed like something new, and then-Governor George Bush turned to the Texas Forest Service to stop what must have appeared to some observers as the natural analogue to an act of terrorism. An organization that had nurtured itself in the piney woods of east Texas prepared, with little additional funding or formal direction, to act as the rural fire department for the second-largest state in the nation.[10]

The buildup came rapidly, a mix of local and national investments with the TFS as broker. The outcome was a distinctively Texan solution that left the state as the primary power.

After the 2006 season, outbreaks paused for several years. In that space the TFS conducted careful studies of the informing "firestorms" to determine why they had been so damaging and how effectively to intervene; strengthened linkages with its federal partners; and crafted a Texas Wildfire Protection Plan in 1998, with legislative funding beginning in 2001. To the question why communities were at risk, the simplest answer was because of how they lived on the land. There were more people living in increasingly combustible places in styles that boosted their vulnerability. The old ways no longer worked since their social and environmental settings had shifted.

Traditional patterns often persisted in such forms as pier-and-beam housing, abundant open space, and small communities thick as mesquite. But pier-and-beam construction allowed embers to catch kindling underneath. The shift from a commodities-based to a service economy transformed what had been a buffer zone around towns into an attractive nuisance for burning. The countryside that had wiped out fire as it once had the bison was sporadically recovering, with significant fractions morphing into brush; and when drought followed a wet year it stoked fires that were to debris burns what a tornado is to a dust devil. A reliance on local volunteers could no longer cope with flames that blasted through communities, not just single dwellings, or surged over tens of thousands of acres, not vacant lots. Nor was it clear that in the new Texas there would be sufficient volunteers to compensate for the Dallas lawyers and Midland bankers who owned the ranchettes that dappled the landscape. All in all, there were lots of contributing causes, but in sum, they pointed to the fact that people lived differently in Texas. The land had regrown, and people had moved out of it in large numbers while simultaneously moving back into it in indefensible ways.

The solution proposed by the TFS argued for a mix of actions. Prevention was the preferred treatment and Texas celebrities (usually athletes or coaches) were enlisted. But fires would happen, and that meant bolstering first responders; and because some initial attacks would undoubtedly fail under the severe conditions that were underwriting the firestorms, it meant creating a reserve force with the capacity to cope with fires over large areas and extended burning periods. To better understand the dynamics of its declared firestorms and to determine when such fires were likely to occur, the agency invested in research and predictive services.

It was a bold plan that provoked Texans to imagine fire protection as more than something staffed by local volunteers and financed by pancake breakfasts. While the Texas Forest Service still employed as its emblem a lone star with a loblolly pine planted in its center, the reality was a five-regioned spread more likely centered on mesquite, oak, or cedar.

The reforms proceeded simultaneously on three levels of government. The biggest push was local, to bolster the VFDs, which remained the first line of defense. As a state forestry program, the TFS had long served as a conduit for the Federal Excess Personal Property program and Department of Defense Firefighter Property Program, which had supplied trucks and tankers often converted to VFD engines or water tenders. This relationship expanded to cover a baker's dozen of federal programs to bolster both volunteer and career fire departments. There were programs to enhance training, upgrade personal protective equipment, buy foam, and acquire insurance for equipment and personnel. The U.S. Fire Administration and FEMA sponsored special grants. After 9/11 there were further programs under Homeland Security (it probably helped that the president was a Texan who as governor had started TFS on its new trail).[11]

To orchestrate VFDs on a larger scale, however, the state itself had to become an active agent. The upshot was the Texas Intrastate Fire Mutual Aid System, analogous to California's state fire plan in that it allowed transfers within the state during emergencies. To stiffen local capacity, TFS acquired and doled out to suitable departments additional engines that were then available for call-up during emergencies (in a process again similar to California's). Access to national fire resources was strengthened in 1998 when the federal agencies established a Texas Interagency Coordination Center at Lufkin, to which the TFS was a partner.

The arrangement left local government as the responsible authority, and the state (through the TFS and the Department of Emergency Management) as the broker between the locals and the feds during emergencies. Although state funding faltered during the recession, when the plan's 10-year sunset clause came up, it was renewed with additional monies. By becoming a matter of public security, the 2011 conflagrations had called the legislature's bluff. Legislators had to ante up or fold. They chose to stay in the game and raised the pot an additional $20 million.

Its late arrival into the national fire scene allowed Texas to learn from everyone else and to tap into national standards and resources. What made its story distinctive is how it reconciled those reforms within Texas culture.

For all the huge expansion in its mission, the TFS has remained astonishingly small, and fire is but one phase of its charter, if a commanding one. In 2011 it had a staff of 447, of which 292 were committed to fire, to oversee 268,820 square miles. By contrast the U.S. Forest Service in Texas had a staff of 200 for 675,807 acres. The major reason for the difference is that TFS was able to work through VFDs, and a bevy of federal programs for equipment, readiness, and emergency response meant the country overall bore much of the costs. (In 2011 the federal government authorized 55 fire-management assistance grants, apart from committing 107 crews, 239 dozers, 954 engines, 246 aircraft, and a cycling of incident management teams drawn from the 50 states and Puerto Rico.) The Texas Wildfire Protection Plan did not have to invent its critical parts. It had only to coordinate them and put their collective efforts where they mattered most.[12]

That meant prevention and suppression. The TFS was charged with fire control, not, like the federal agencies, with fire management. Its goal was public safety, not land renewal or ecological integrity. It did not administer the public lands of Texas for multiple purposes for many clienteles: it sought to control wildfire and harden assets to prevent loss of life and property. It had to do one job well. Revealingly, its interest in mitigation relied heavily on Firewise and fire prevention, not on fuel management. People would accept messages, particularly from celebrity Texans. They were not likely to accept prescriptions on what they did with their land. That sounded like profaning the sanctity of private property. The solution crafted by the Texas Forest Service is probably the best (maybe only) one possible.

The state needed an innovative institution, adapted out of southern roots and lean but effective, to deal with outbreaks of violence that threatened settlement west of the Balcones escarpment. As it had done in the past, it took an idea from the woody east, mounted it, and sent it to the grassy frontier. In a famous study Walter Prescott Webb quoted admiringly the old description of a Texas Ranger as a man who "can ride like a Mexican, trail like an Indian, shoot like a Tennessean, and fight like a very devil!" That was old Texas. In new Texas TFS staffers had to research like a professor, negotiate like a lawyer, talk like a country singer, and fight some very devilish fires. What endured was the need to adapt locally to the two very different realms of Texas, one humid and one arid, one forested and one grassed, one ending the South and one beginning the West.[13]

What Texas invents, particularly institutions, tends not to travel well outside of Texas. For all its Hollywood sound and fury the Texas livestock industry fared badly beyond the state. Its droving era passed quickly, and the laissez-faire herding which abandoned stock to fend for themselves left longhorns to die by the thousands in northern blizzards and ranchers unable to compete with better breeding and husbandry. The extension of the practice into New Mexico and Arizona helped drive fire out of public lands as it did in the open range and private ranches of Texas. The scars from that stampeding era still linger.

In the case of its statewide fire program, however, Texas doesn't have to emigrate, only assimilate. It remains for America and Texas to reconcile their relative contributions, which will likely again leave Texas with responsibility for its own lands, which will include sotol as well as shortleaf, and winter grass as well as hardwood rough.

SIX FIRES OVER TEXAS: A RECONNAISSANCE

SOUTHERN STRATEGY (EAST TEXAS)

Begin in the east, where Texas as a polity began. The Sam Houston National Forest lies between a giant statue of Houston himself and San Jacinto, where he won Texas independence. But the piney woods of east Texas are also where Texas forestry began, and hence the state's hearth for fire protection.

The Texas story is essentially continuous with the rest of the Southeast. By the late 19th century the region's longleaf and shortleaf were being rapidly cut out for a national market; Texas joined that vast slashing. When the state authorized a state forestry program its woods were fast going (or gone) and the Texas Forest Service was charged with bringing the carcass back to life. The primary goal was to reforest, which, in the understanding of the day, meant replanting and fire control. In the early

Depression, large swaths of cutover lands fell to tax delinquency and were abandoned, and in 1934 were acquired for a cluster of four national, and a scattering of state, forests. That established a formal point of entry for state-sponsored conservation to operate on the ground.

Fire protection was a core doctrine: it was the strong nuclear bond between state and federal forestry. Its ecological continuities and common settlement history soft-welded Texas to the rest of southern forestry; the Weeks Act and Clarke-McNary Program joined the TFS to a national network; and the acquisition of federal land for forests, parks, and wildlife refuges put federal boots on the ground. Texas foresters handled fire as their southern colleagues did. At first they attempted to exclude it to encourage regeneration, then they selectively adapted prescribed fire, and eventually they made wholesale burning a foundational practice. National forest headquarters for the state is sited in Lufkin. The forestry school is housed at Stephen A. Austin State University in Nacogdoches. Texas belongs to the Southern Group of State Foresters, the Southern Forest Products Association, and the Southern Pine Council. While Texas forestry speaks with a hardier twang, its fundamental accent is southern.

All of these relationships, however, thinned, and most broke, as Texas sprawled west. That drying frontier spawned ecological outliers. The Lost Pines around Bastrop are a Pleistocene relic that survived by special adaptations. The red-cockaded woodpecker, an endangered species understood as inextricably identified with longleaf pine, here colonized loblolly pines instead. Commercial forestry faltered as sawmills remained to the east and oak-dominated hardwoods and brush characterized the central woodlands. Dryland farms yielded to irrigation, and logging to ranching. Social arrangements followed suit. The federal holdings did not extend westward; and Texas's own land agencies did not forge compacts with neighbors. As they moved west, fire institutions faded and diminished, like lofty loblolly yielding to mesquite and then to scattered tobosa. The southern model couldn't pass over the plains. Fire practices crafted in pineries withered amid bunchgrass. Prescribed fire for fuel reduction worked in woods, but not in pasture.

Still, the system succeeded in its hearth. When the 2011 firestorms blew across Texas like tumbleweeds, the Bearing fire burned 20,800 acres and went to ground on "the Davy" when it hit patches previously prescribe burned. That made it the largest recorded fire in east Texas history;

though amid 3.9 million acres, it was twice two orders of magnitude less than the state's reckoning for the year. The Sam Houston prescribe burned that much annually, and was approaching 40,000 acres a year.

The two Houstons are a study in contrast, of which city and forest, separated by 50 miles and 7 million people, are only a point of departure.

Houston the city has an economy predicated on fossil biomass and an urban ecology powered by industrial combustion. Houston the forest has a natural economy based on a loblolly-shortleaf forest sustained by open burning. Houston the city famously has no zoning and allows its geography to be inscribed by the sprawl of free-ranging automobiles that have put it out of compliance with Environmental Protection Agency air-quality standards. Houston the forest has an intricate breakdown of burn blocks, each adapted to the varied uses of the land and all subject to vigorous public discussion under the National Environmental Policy Act.

City and forest join in ways far more complex and nuanced than the threadbare expression "wildland-urban interface" suggests. The city relies on the forest for recreation and ecological services. The woods, in fact, are a magnet for exurban outmigration. The urban economy directly, and its indirect questing and valuing of natural amenities, far outweighs the old sawtimber cash crop of the pineries. Private landowners with inholdings among the national forests are replacing pine plantations with subdivisions. But the city can influence land use indirectly as well, by extending values that promote hiking or wilderness over grazing, for example. That urban populations particularly value endangered species grants the forest leverage through the presence of the red-cockaded woodpecker. Urbanites are more likely to tolerate or even promote a federal presence, as happened in the 1970s when patches of residual wildlands were stitched loosely together to make Big Thicket National Preserve. In such ways, the city sustains the forest.

Yet the city also compromises the capacity of the forest to thrive. The national forest is a quilt of public and private lands, with the private patches evolving out of the kind of rural usage that tolerated fire and into urban ones that don't. The city especially detests smoke—it has too many emissions of its own, including invisible effluents like ozone. This means that the forest can't burn in the traditional southern way with a north wind from a passing cold front behind it. That urbanites spend leisure time in the forest more or less year-round makes the scheduling of burns tricky; there is always a cyclist, hunter, hiker, or scout troop within a burn unit. Nor does "the Sam" believe it can burn into the growing season; the peculiar adaptations of its flora to more western conditions nudge its fire ecology slightly out of kilter with regional norms. The forest features enough oddities that the southern strategy won't transplant directly, yet not enough to qualify the woods as something unique in themselves. Along this frontier of fire, as along frontiers generally, survival depends on adaptation.

In ways typical of the Southeast, but on a scale unimaginable in western wildlands, the Sam burns. Probably, on average, it burns half the acreage of what a southern burner would consider ideal. This is still an order of magnitude greater than western counterparts achieve. Its fringe of loblolly is a kind of biotic Hadrian's Wall that traces the limits of the southern model. The gradual removal of grazing allotments has further shifted the burden of fire management, which is to say, fuel management, onto prescribed burning. Probably, only the weight of federal resources and legal obligations allows even this scale of operations to continue. Farther west they stumble, and beyond the 100th meridian, they fall into virtual ditches of despondency.

As the Sam shows, there is another gradient to the Texas fire scene than the one that runs east to west. Superficially, it appears to run south to north, from city to forest. In fact, it runs down to up, from buried biomass to surface combustibles. The border between the two Houstons ultimately traces an exchange between two realms of combustion. The geography of the WUI and the rest is just an outward manifestation, like an exoskeleton giving it shape. In this regard, east Texas is not only where the Texas fire story begins but also where it is trending. The two Houstons exhibit in compressed form the longer and wider story of fire across the state.

HIGH PLAINS DRIFT (NORTH TEXAS)

American historiography also has its dryline. Since John Wesley Powell published his justly celebrated *Arid Lands* report in 1878, the line at which annual rainfall drops below the level needed for dry farming has been taken as the divide between eastern and western landscapes. The grasslands appeared earlier and demanded a variety of innovations for fencing, fuel, plowing, and shelter; but not until a stubborn aridity denied the old crops—roughly along the 98th meridian (usually rounded to 100th)—did western settlement stumble and halt. East of that line trees could grow if fire was kept out. West of the line trees too faltered, and grasses replaced them. Crops required irrigating, and herding made more sense than farming. This climatic divide became a settlement divide; it is a major reason the bulk of public lands resides in the West. The line has traced a meridian of historiography as well.

What is true for the plains generally is certainly true for Texas as well (this is the baseline thesis for Walter Prescott Webb's *The Great Plains*). The dryline meridian arcs along the Balcones fault line, and it defined the line of settlement under a broadly southern model until after the Civil War. Before then aridity, hostile Apaches, Kiowas, and Comanches, and a sparse population made the arid plains the frontier. One might also add wildfire to those settlement inhibitors since it was as endemic to the region as prairie dogs and grama grass and was capable of propagating at rates and breadths unknown in the piney woods.

The fires burned widely and routinely in an immense choreography between climate, grass, a massive menagerie of grazers, and fire, all loosely staged by the alpha creatures, humans, who also happened to hold a species monopoly over ignition. The details are unclear and were, in any event, still adjusting since the adoption of the horse by plains peoples less than a century before and the resulting internecine conflict over territory. But there is little reason to doubt that similar relationships to what Surveyor-General Major Thomas Mitchell wrote concerning Australia at the same time also held for fire, grass, and bison, and human inhabitants on the southern plains.

> Fire, grass, and kangaroos, and human inhabitants seem all dependent on each other for existence in Australia; for any one of these being wanting, the others could no longer continue. Fire is necessary to burn the grass, and form those open forests, in which we find the large forest-kangaroo; the native applies that fire to the grass at certain seasons, in order that a young green crop may subsequently spring up, and so attract and enable him to kill or take the kangaroo with nets. In summer, the burning of long

> grass also discloses vermin, birds' nests, etc., on which the females and children, who chiefly burn the grass, feed.

Unsurprisingly, America's aboriginal peoples referred to free-ranging fire as the "red buffalo."[14]

The longer the fetch of grass and wind, the vaster the fire. Accounts from the late 19th century report burns of several million acres. By then the old choreography was breaking down as the mounted aborigines were swept away, the bison driven to extinction, routine fire setting banned, and the geography of combustibles fractured into fields, pastures, and roads, and no longer joined by subcontinental migrations of grazers, hunters, and herders. But where those elements endured, or were recreated, there might be nothing to stop a running fire fed by grasses from a wet rainy season other than wide rivers, swathes eaten out or trampled by bison throngs, the occasional arroyo, or a soil-cracking drought that shriveled the combustible grasses before they could carry flame.

After the Civil War the rush of settlement resumed. It boiled out primarily from South Texas through the Hill Country as hybrid cattle and hybrid herding, a mestizo fusion of southern and Mexican practices, pushed into fresh territory. The process assumed the form of a colossal swap as Texans replaced Comanche; longhorns, bison; and cattle drives northward, the seasonal migration of the southern herds. The upshot was a less controlled mix of fires—a fast-morphing regime—but one that still left fire on the landscape. A prairie fire remained a threat; but fire was also recognized, if timed properly, as a means to green up forage and retard wildfire.

The era lasted only a couple of decades. The fertile sites were gradually plowed into cotton and alfalfa. The open range was fenced into pastures. The grass was grazed to nubbin. The longhorns were left feral in remote canyons while improved breeding replaced droving with something like husbandry. But the droving era left behind more than popular songs and Hollywood celluloid. It repeated the tragedy of all pioneering: it changed the conditions that had made it possible. Its creatures, its lands, the ecological choreography that organized them into patterns—all went.

The spoor of the droving era contained the seeds, literally, of a new landscape. The cattle trails tracked what Texans came to call "brush"—cedar around the Hill Country, juniper to the west, creosote in the south, and mesquite generally but especially to the north. Eating mesquite seeds, scarifying them through the digestive track, and then defecating them in manure piles atop the trampled soil created an ideal scenario to propagate mesquite woodland from small enclaves onto vast swathes of north, central, and west Texas. The woody species are several, but all are sensitive to fire in their initial year or two, and some, like redberry juniper, remain vulnerable for up to eight years. Likely the trees had always been around, though only in fire-free niches among rocky outcrops and east-side riparian patches (downwind of barriers). The cattle carried them north and west, not only spreading seed but cropping away the fires that might flush seedlings from the scene. As fire became rarer, surviving brush became more common.[15]

Once rooted, fire alone could remove only a few species. While a stiff burn might top kill a fraction of the others, primarily sprouters, it could not uproot them, and they regrew like hydras. Mesquite, in particular, thrived. It naturalized among the new fire regime—one it had helped create. Once they achieved a critical mass, the trees could self-propagate. They spread like a slow plague. What had been identified at the time of European contact as grasslands shape-shifted into brushlands. A new, self-perpetuating community of trees, grasses, browsers, and grazers coalesced; what cattle had begun, other seed feeders continued. Unless cut back, poisoned, or uprooted, the brush remade the grassland into a savanna, and then into a woods. Texas ranchers have been trying ever since to expunge them, without success.[16]

Most of what is known about fire ecology in Texas comes from elsewhere. Its eastern forests are studied as a subfield of the southern pineries. Its grasslands are mostly known from research in Kansas and Oklahoma, and this includes the dynamics of encroachment by eastern red cedar. Its coastal marshlands extend those that fringe the well-investigated Gulf. Even the Chihuahuan Desert is better researched in New Mexico. But in one area Texas stands nearly alone. It is a world center for the relationship between fire and the varieties of brush, particularly mesquite, that have come to infest the Texas landscape.

Fire and mesquite establishment, fire and mesquite-grass dynamics, fire and grazed pastures in mesquite woodlands, fire and mesquite control or outright eradication, and in recent years fire and what might be termed the mesquite-urban interface—these have claimed the special attention of fire science in Texas and produced a world-class literature. The what, why, and how of this research program speak volumes about the peculiar character of fire and its management under the flag of the Lone Star state.

The research might be better characterized as fire agronomy than fire ecology in that its funded purpose is not to advance fire science fundamentals so much as to find ways to control or kill off mesquite and thus improve pasture. The reductionist tendency of science lends itself perfectly to such purposes: it can isolate particular features and find ways to manipulate them in order to maximize productivity. The ultimate outcome is not reckoned in ecological integrity or grassland restoration (new grasses may be introduced) or some other public good, but in economic wealth. Sponsors hope fire will prove a cheap tool for brush clearing and that restored pasture will yield better fodder for more beef.

Mesquite is stubborn. If you want to remove it you treat it as Rome did Carthage: burn, plow, and sow with salt. Or in modern parlance mechanically chop or (preferably) uproot it, spray it with herbicides, and, when conditions permit, burn it out. Unfortunately, while fire is the most benign and cheapest method, by itself it is the least effective. Like prairie bunchgrass mesquite has deep roots, resprouts readily, and thrives when burned. Maybe a millennial drought and fire could deliver a one-two punch. Otherwise, what a killing fire might do is instead delivered by machines powered by fossil fuels or poisons extracted from fossil biomass. Both are expensive and require maintenance regimes. An alternative strategy might be to harvest the wood and burn it for bioenergy, which would help cover the cost of clearing. Economics and ecology are out of sync because, in such a landscape, fire is a process for renewing rather than replacing the biota.

The most likely outcome is an accommodation. Accept that mesquite has naturalized, like armadillos. Control it so that useful grasses can still flourish and feed herds. Unlike juniper mesquite does not snuff out grass, nor does it, as juniper can, convert surface fires into convection-dominated conflagrations. Rather it shifts and mixes species, lessens the density of grasses, and complicates working with stock, but it can be lived with. Where it becomes too encumbering remove it by means that pay for themselves. And where possible use fire to refresh the landscape and encourage healthier grasses that can compete with brush.

Yet there is more to the story of mesquite as a parable of Texas fire. The grasses that thrive in such soils and climates evolved with and still flourish with fires that come amid long growing seasons and hot summers. The system broke down because fire was removed, and it is breaking down again because fire is returning in feral avatars. After being buried for so long, what accounts now for its zombie reemergence?

This is a region, as Jim Ansley notes, ever latent with fire, always on the cusp of an outbreak. Its tenacious grasses, its blustery weather, its knife-edge teetering between economics and ecology that determines how many cattle or deer are on the land and how many volunteers are available for a callout—any can cause the landscape to veer into fire. So far the concern has focused on wildfire, and, apart from the usual suspects for causes, observers point to reduced stocking because of persistent drought, a shift to hunting income in place of cattle sales, fallowing as a result of the federal Conservation Reserve Program, conversion to exurbs and ranchettes, and reductions in state support for VFDs, all of them economic considerations that translate into more combustibles. It's a slow drift that is enough to tip the scene into fire.[17]

Behind such tinkerings, each minor but perhaps enough to trigger a big response, like a switch that turns on a dynamo, lies a larger notion that seems particularly suited to Texas's obsession with its landed estate. Fire, grazing, and grass work when there is abundant room, when there is ample space for land to burn, grass to grow, and land to lie fallow. Herds move between the black and the green, and big herds and arid landscapes require lots of land to accommodate this migration. Settlement first replaced bison with longhorns, then shattered that mobility by closing the range. Ownership—the sanctity of private land—stayed within fences, and even the largest ranches barely had enough land to support the level of cattle they needed to generate the income they wanted. The smaller ones had to move additional feed to the herds, or ship the herds elsewhere for a while. A basic principle of fire ecology is that fire recycles. In the case of grazing economies the animals must cycle, too. If wildfire takes out a significant fraction of winter range the ranch will suffer and perhaps crash. The usual Texas solution, substituting more land for intensive management, can't apply.

An obvious solution is to burn off that surplus under controlled conditions, which would reduce the risk of wildfire and enhance the long-suffering grasses. But fire management also requires intellectual and political space. Fire must be seen not simply as a threat or a tool but as a useful process. The tendency in Texas, however, has been to define it as bad because it threatens property and removes forage. Instead, to prevent fire and promote beef, the default strategy has been to graze down to the point that fire can't spread. In a sense, settlers removed the grasses that sustained fires as they did the bison that supported Plains tribes. While the productivity of the soil, of the system overall, may degrade, and may promote yet more brush, the immediate outcome is favorable, and in a landscape locked into boxes of individual lots, that is a rational economic choice that defers costs to future generations or to the public.

The paradox of course is that the same processes that are restoring bad fire underwrite the possible reintroduction of good fire. At core it requires surplus combustibles, which is to say, fallow, or land out of immediate production. While the long-term benefits might be healthy, it is the short-term cost that dominates the calculus of any but the largest landholder. County judges and commissioners are quick to declare burn bans. Many ranchers, particularly in the West, don't want fire of any kind, except where it might remove brush, which in practical terms means combustion housed within engines that crush, shred, or extirpate. To date, the recovered and remade landscapes are fueling wildfire. To what extent prescribed burns might substitute will decide what kind of fire regime will prevail in the future.

Public lands, or private lands used for public good, can make that transition easier because they can provide the requisite scale of operations, and they can absorb or internalize those immediate losses for longer term benefits. They don't have to mark ecological goods and services to market; they can invest at a social level for social amenities. They have a public purpose in their charter. Those lands needn't be federal. Texas has a robust complex of state parks that boast integrated fire-management programs, complete with prescribed burning. The tide doesn't have to rise high, but something needs to float land use above the quotidian mean if it is to have enough political and ecological space in which to insert controlled fire.

WEST OF THE PECOS (WEST TEXAS)

Stephen Crane once complained of a novel that it "went on and on, like Texas." The Republic of Texas went all the way to the Rio Grande and up into Wyoming. In its final shape it surrendered Santa Fe but held on to El Paso. That brought under its nominal control all the land west of the Pecos. The pattern of settlement that had crossed the Sabine now stretched so far west that it was no more than a thread.

The lengthening seemed to pull the land to the point that it collapsed into a patchy geography of mountains and sun-fried plains. Piney woods and southern rough shrank into pockets of pinyon, juniper, and the scruffy sotols and bunchgrasses of the Chihuahuan Desert. Armadillos replaced possums. A labor force of African Americans yielded to one of Hispanics. What happened with natural history happened with institutions. A fire-protection system based on relatively dense settlement extenuated into VFDs, as scattered as mottes, and then into private ranches. Counties or big landowners replaced even the state.

The upshot was that Texas held a landscape that bonded it to the Southwest, a region that had trekked beyond the laws of its settlement history. It was the kind of landscape that farther west the federal government oversaw, but because it was in Texas the state had formal responsibility without any effective mechanism of control. West of the Pecos, county judges and big landowners ruled. Natural history bonded the region to the west, and human history to the east. Fire protection derived from laissez-faire overgrazing. Standards for practice were "what my daddy did."

The trans-Pecos region is different, the one geomorphic province of Texas that is not part of the Great Plains. Geologically it resides with the

Basin Range province, and ecologically, with the Madrean sky islands. It belongs to Texas politically only through a freak of history, as the rump of a westering imperial Texas. But that exception makes it a fire province all its own.

The Davis, the Del Norte, the Eagle, the Delaware, the Chisos Mountains—all resemble the constellation of peaks that extends west through northern Mexico and the southern American Southwest, and like those in Arizona even have an observatory (McDonald). Compared with the others, the trans-Pecos peaks are more muted and less watered; but their big difference lies in the history of settlement and governance. Excepting Big Bend and Guadalupe Mountain National Parks, both of which were acquired long after settlement, the region has no serious federal presence. The original public estate was sold or bestowed in grants.

These differences express themselves in fire as in other matters. The trans-Pecos fire regimes are fundamentally akin to those across the Madrean Archipelago, with a significant fraction of fires started by lightning. The region seems especially susceptible to dry lightning—its terrain is high enough to force thunderstorms and low enough to have rainfall evaporate into virga. Bolts strike small trees or grasses. In 2011 the Van Horn District had 21 fires, of which 13 were kindled by lightning; the Fort Stockton District had 106 fires, 55 from lightning. The TFS West Branch operations center overall experienced 221 fires, 48 percent from lightning. Fires will start. The issue is whether they spread, what benefits or damages they caused, and what it takes to stop those that are not wanted.

What shaped the fire dynamics of the region's natural history was human occupation. Settlement history west of the Pecos was typical of the Southwest generally. It meant mining, logging in the mountains, and grazing from mixed herds of livestock everywhere—grazing intense enough to wipe fire out of the landscape. It doesn't take much grass to carry fire, but where grass is the primary fuel, the high-intensity fires, like those throughout the Southwest generally, tend to follow a succession of wet years that can grow fuels faster than ranchers can build up herds to crop them before dry years return and fire claims the remainder. What has changed the dynamic in west Texas is that land use, or economic cycles, and changing politics now syncopate with climate. Prior to 2011 the worst drought of the century occurred in 1956. Across that part of Texas under TFS responsibility some 4,648 fires burned 76,171 acres. Undoubtedly there were more fires and acres, but interest did not go beyond local landowners. Amid the 2011 drought 29,394 fires scorched 3,933,436 acres and wrought far more damage. The fires were worse, they were recorded, and their impact ranged beyond strictly local concerns. The worst drought in 55 years had more to act upon—more and different combustibles, landscapes organized to diverse purposes, a hierarchy of institutional responders. What changed was not simply the fire numbers and acres burned, but their social setting and consequences.[18]

Slowly, patchily, the ruinous grazing that had stripped fire out of the region had receded as a result of its own excesses, the removal of government subsidies, and changes in ownership that replaced grazing with other purposes. In one spectacular case, the Big Bend region, land passed from private to public hands. In 1933 the State of Texas acquired the land to make Big Bend a state park, and with subsequent purchase by the federal government, it became a national park in 1944. Even with formal protection by the feds the landscape took decades to recover. Probably the only fires anyone heard about were those that kept pots simmering at the annual Terlingua Chili Cookoff. Big Bend, in fact, illustrates the extraordinary isolation of the region on the American fire scene. Not until 1996 was Big Bend granted a fire-management officer. If fires occurred (and most were still small), the park had to look to neighboring ranchers for assistance, in effect behaving like a giant dude ranch itself. When a fire bust broke out in 1990, with many starts but not enough acres to command national resources, the park turned to neighbors across the Rio Grande and, in what resembled a trans-Pecos version of the Southwest Forest Fire Fighters, with later approval of the Department of Homeland Security and the U.S. Border Patrol, created a fire crew from willing workers in Boquillas, Mexico. Eventually the program expanded to three villages, and the Los Diablos crew moved beyond Big Bend.[19]

So the federal presence was miniscule. Meanwhile, although agriculture still dominated the regional economy, ranchers began pulling back or just sold out. The mountain regions became the site for recreational facilities or homes. Absentee ownership became common, although sometimes remaining within a family as generations moved from pioneer rancher to college-educated son to banker grandson, who held to it as a family heirloom, like British aristocrats with ancestral mansions, leaving the lands no longer chewed to dust in a search for maximum beef, mutton, or wool. Jeff Bezos bought a million acres to establish a spaceport for Blue Origin. Even if the private holdings were large, however, they represented a fragmentation in use that created room for fire. Amid the cycle of drought, producers sold off stock rather than pay for imported feed, which further lightened the weight of grazing. In this new west Texas more grass was available amid droughts, and west Texas brush could thicken and spread. The fires mutated from small or blitzing grassfires into larger and more plume-dominated conflagrations; and some became monsters. The 1993 fires drew the Texas Forest Service into the region for the first time. The 1996 Big Country fire, along with those in north Texas, was the catalyst for expanding the writ of the TFS across the state, and through the TFS for integrating the trans-Pecos into the national economy of fire. Thereafter, almost every other year, west Texas blew up. With the 2002 season it began to command national resources to fight back the blazes.

The institutional landscape struggled to catch up with the scale of burning. On the remote holdings landowners remained responsible, and near small towns VFDs continued as the first line of defense. The new regime, however, could quickly overwhelm them. They needed air tankers, not cattle sprayers mounted on pickups, and mechanized task forces, not cowboys with swatters galloping around flames like a roundup. In 2011 a single lightning bust started 23 fires, of which five exceeded 23,000 acres. The locals had to turn to the TFS, whose main west Texas cache was in Midland—a couple hundred miles away. The fires were, as locals note, "fast and furious." By the time reinforcements might arrive the fires had either expired or blown over the next ridge. The Texas Intrastate Fire Management System could call up yet more engines and planes, but national resources were frequently closer. One way or another, the outcome often meant an appeal for federal assistance.

National integration had its own instabilities. From a countrywide perspective west Texas (all of Texas west of the 100th meridian) looks as though it belongs under the Southwest Interagency Coordination Center in Albuquerque, which is closest and has similar kinds of fire, while the rest of Texas aligns with the Southern Interagency Coordination Center in Atlanta. Among states only Idaho is similarly divided. But 63 percent of Idaho is federal land (and almost all private land lies to the south); less than 2 percent of Texas is. The medium that holds the state together is the state.

Over the past 15 years much of Texas has worked out how to use the TFS as a broker between local interests and national resources. That process, in accordance with much of Texas history, has extenuated the farther west it has gone. Across the Pecos it entered new country, what must have seemed politically at times like the tribal lands of Pakistan.

In the 2011 outbreaks the calls for assistance went first to the TFS. For the mammoth Rock House fire that bolted from an abandoned house outside Marfa across the Davis Mountains, the call then went to Atlanta. The incident management teams who arrived had little experience with either trans-Pecos fires or politics, with grassfires that romped across 30 miles in a single burning period or with county judges deemed almighty in their baronies. They met a community that could not handle the crisis on their own but were reluctant to relinquish control. These were, as one observer noted, folks who were furious that Google Earth had posted images of their spreads and were not about to let an outsider decide when and where to cut fences, scrape lines, or burn out. Local volunteers showed up in normal work clothes and pickups, determined to do something but hesitant about submitting to national standards for personal protective equipment, pack tests, or S190. They showed up; they dispatched themselves. They seemingly fought stray flames as they might round up mavericks. That is what they had done, more or less successfully, for over a century. But these were fires of a ferocity no one had seen before.

The national fire teams were accustomed to operating on public lands or lands that subordinated private ownership, during emergency, to a common good. West Texans were not. The postfire review has been prolonged.

THE BURNING BRUSH (CENTRAL TEXAS)

The Edwards Plateau spreads like a shallow dome over the center of Texas. It isn't high, so its valleys and ravines aren't deep, more like a gentle scalloping, but it introduces some rumpled relief to the landscape, a terrain that seems to blend in a subdued way both western mountains and coastal plains, and has long had a similar medley of grass and brush. Here, too, settlement traditions from east and south mixed and mingled before pushing west and north. In the historical geography of Texas, the plateau forms a high ground.

What matters for fire history is that the plateau likewise roughens the pattern of land use, yielding a variability not present elsewhere. The land was too rocky to plow, so like other Great Plains sites that resisted conversion to farming, it turned to grazing, but with an Edwards twist. Here, American South, European, and Mexican herding converged such that pastoralism mingled cattle, sheep, and goats along with the ubiquitous horse. As grazing moved west and north, it simplified; but on the Edwards it retained its mixed stock (the plateau remains a premier producer of wool and mohair). Even its brush stirred together live oak, mesquite, shrubs, and the several species of juniper invariably known locally as cedar. And, as if in emulation, that tendency toward working diversity even expressed itself institutionally. Here private and public interests have found ways to work together not just to fight fire but to reinstate it. The burning brush of the Sonora Research Station has became an oracle for fire management throughout the new Texas.

In broad terms the story is the familiar one. What had been a southern-plains Serengeti fed by grasses renewed by recurring fire shattered from overgrazing and underburning. In many respects, the scenario that has characterized the fracturing of western firescapes—cutting, overgrazing, fire exclusion—commenced here. Through the cracks in the prairie grew woody shrubs and drought-toughened trees. With nothing to drive them out, the invading woods replaced the grasses and rendered the pastures uninhabitable. Once established the brush proved onerous and costly to remove. Even with mechanical destruction it could wholly reclaim a site within 30 years.[20]

What makes the Edwards story fascinating are the details. The plateau has a limestone cap, which makes for shallow soils and undependable surface water. The solution was to found pastoralism on sheep and goats, which migrated from Mexico and were tended by Mexican shepherds.

The flocks' seasonal mobility obeyed the rhythms of water and fire. In the dry season the sheep moved to the permanent rivers, and in the wet, to the uplands in what must have seemed a miniature *transhumancia*, a seasonal migration between pastures. The most ferocious burns occurred in August, which was unsurprisingly the time of maximum grass productivity and occasional dry lightning. But they were likely set anytime the range could accept them. If the transhumance model had in fact been adapted shepherds would have burned behind the moving flocks. Other newcomers learned about burning from the remaining indigenes; as late as 1921 old residents recalled how the natives had "habitually burned off the grass in the spring and fall preceding the rains to keep down the underbrush, to provide green grazing for game and their ponies and to improve hunting." Between them frequent fire and browsing held any brush in check.[21]

Brush there was—it's more indigenous than meat goats, Rambouillet sheep, or for that matter, horses. But fires held it in incombustible niches like rocky ledges or riparian gulleys. The new regime, however, could overgraze in ways not possible for migratory bison and resident whitetail deer, and fire could not compete with those hordes for fodder. The scene worsened when, with the invention of windmills, grazers could remain year round; when with permanent water, cattle could be added to the herds; and when, with barbed wire, private lands shut down even seasonal movement. The herds stayed, trampled, and ate. Grasses decayed, or just disappeared. Without fire to sweep them aside the woody brush thickened. Cutting for fuelwood or posts could wreck the structure of those woodlands but could not compensate for their subsequent viral recruitment in the absence of flame.

The new regime drove out fire as it did the Apache. The newcomers' fear of prairie fire fused with their desire to maximize meat and fiber and led to a common solution: heavy grazing. To the practical logic of settlement, which was stripping fire away, herders added legal threats. As early as the 1840s laws appeared to regulate burning. In 1884 the Texas legislature made burning range a felony. The more fire remained off the landscape, the more degraded the habitat became, and the less productive for anything save brush. What seemed a minor ecological infection as cedars began to propagate outward from their enclaves became an all-out plague. A semiarid savanna evolved into a brush field jungle. Further squeezed by that shrinking range ranchers stocked as heavily as possible. The Edwards Plateau ecosystem spiraled downward.

So far, so common. The same story was playing out in central and south Texas that defined nearly everywhere else. What makes the Edwards story vital is that the circumstances have evolved, as they failed to do elsewhere, an institutional response that found a common public and private cause.

Stung by a decaying range and ill animals the regional ranchers organized into the Texas Sheep and Goat Raisers Association. At its 1915 inaugural convention in Del Rio, the association sought to establish a research station to bring modern science to bear on their concerns. They enlisted Texas A&M and the state legislature for help, then contributed half the cost of a 3,462-acre ranch between Sonora and Rockspring as an experimental ranch. Substation Number 14, later known as the Ranch Experiment Station, and still later as the Sonora Experiment Station, was established in 1916.

The station took as its charge research into breeding, nutrition, diseases, poisonous plants, and pasture rotations. But clearly the stocks' internal health depended on the health of their external environment. What drove the awful economics of ranching was the de facto implosion of the range. There were too many mouths for too little grass. What biomass thrived, the cedar, was unpalatable—in fact, it was toxic with terpenes. When it reached 30 percent canopy cover, it effectively voided pasture. From time to time chaining or bulldozers would crush and sweep the hated brush from the scene, only to watch it return. Within 20 years the brush dominated. After 30 it again rendered the landscape economically uninhabitable.

What the two economies, nature's and humanity's, needed was a cheap, benign process other than driving over the landscape like a monster truck rally. What it needed was fire, not only as a means to burn woody fuel but as an all-purpose ecological catalyst, a enabler for the parts of the system to begin acting in concert and turn the sinking gyre into a upward spiral. This, however, flew in the face of implacable fears, learned traditions among the folk, and, not least, the superstitions of the learned. Rangeland texts still railed against fire. Government-sponsored conservation distrusted fire and discouraged its use. Agricultural subsidies encouraged grazing as close to the margin as possible, for which fire was neither desired nor possible. Since 1985 counties declared burn bans in what seemed to critics like contagion. No one, it seemed, wanted fire; but fire was the one thing that could resuscitate the system, could administer ecological CPR to a dying landscape. The only way to reform the habitat was to reform the institutional context that determined what happened (or didn't).

In fact a few niche ranches had kept burning alive, and as prescribed fire became a common theme nationally during the 1970s and 1980s, government institutions (the State of Texas mostly) offered some assistance as part of an extension program. But although the agencies might accept fire in principle they still suspected it in practice, and prescriptions tended to be so conservative that the fires neither escaped (as feared) nor did the work needed (as desired). It fell to the Sonora Research Station to find the right compromises and to Butch Taylor to broker an outcome.

Charles A. "Butch" Taylor had grown up in a west Texas landscape in which he saw few free-burning fires. No one burned deliberately, and wildfires were rare because every place was grazed to a buzz cut. When he entered the army, however, after a degree in range science, he saw frequent fires at all seasons at Fort Sill, Oklahoma. Ordnance constantly started fires in mixed grass, some of which burned fiercely. The prairie that grew amid the burns was the best he had ever seen. At Fort Hood he watched a grassland savanna flourish amid tanks and shellfire. In Vietnam he saw what should be, according to climatic factors, a tropical forest grow house-high grasses under a regimen of annual firing. What any reasonable observer might classify as hostile fire was producing grasslands better than any he had encountered under the management prescriptions favored by officials and academics. Somehow, he knew, he had to get fire back into the Edwards Plateau in a big way.

He experimented; he mixed regimes and practices. Fire alone couldn't do the job, but fire interacting with other techniques could, and once a site was converted from brush fire could maintain it. Surprisingly the hotter the fire the better. The landscape was almost catatonic: it needed shock therapy. Fortunately Texas grasses are tough, robust, resilient; given breathing room and rain, as little as two good seasons out of five, they can recover. In 1997 Butch staged a tour for the region's ranchers in which he showed a series of treatments. All began with mechanical clearing, subsequently subjected to different burning regimes. The more and hotter the fire, the better the outcome looked to the ranchers. That explained the what and the why of burning.

The problem remained the how. There was no obvious point of leadership, no state agency with the resources or commitment, no federal bureau with which to cooperate. Given Texas land ownership—even the Sonora Research Station was half-owned by ranchers—the only solution was to build an institution among the local community. The upshot was a nonprofit corporation. The Edwards Plateau Prescribed Burning Association (EPPBA) sought to pool resources, conduct training, reeducate ranchers into the lost culture of fire, and reverse the social and political antagonism against burning. It was neighbor helping neighbor, and once neighbors joined, their individual liabilities shrank because they were a collective.

The EPPBA grew. By 2011 it had spread across 15 counties, each with its own chapter. It was, in truth, the flip side of the VFDs that staffed the fight against wildfire; it had to be if it was to thrive within the political matrix of Texas. As the Texas Forest Service was doing with the VFDs—building up their capacity, imposing some standards, providing insurance—so the EPPBA did for its members. Before long, associations began to spring up in other regions—enough that a Texas Alliance of Prescribed Burn Associations was congealing. That, in turn, could ally with the national Coalition of Prescribed Fire Councils. At county, state, and national levels an infrastructure had emerged. The EPPBA found a way to leverage fire management beyond the isolated landowner and single-purpose productivity. The geomorphic and ecological diversity that characterized the Edwards had found an institutional expression.[22]

The outcome, to date, is more symbolic than transformative. The actual area involved is small; probably land converted to second homes is propagating faster than the acres treated by fire, which is to say, lands that will never burn deliberately are spreading faster than those that might. But like those protected enclaves of cedar from which the brush was able to take over the countryside, the prescribed burn associations are a point of entry for what partisans hope will become a therapeutic infection. For good or ill, fire propagates. The EPPBA wants to grant it room to spread. The association's motto: "Happiness is smoke on the horizon!"

The reality is that it took more than fire's extinction to trash the landscape, and it will take more than fire's return to rehabilitate it. The pressures against fire, and against a turn toward public goods, remain powerful. The EPPBA and its confederates will need tools to push back against the brush encroachment of social and political realities.

A beginning is to leverage fire's ecological power—make it more effective, grant it a wider reach. A successful fire regime will need to be mixed. It will have some hot fires to jolt the land out of its woody coma, some cooler or dormant season burns to prevent reencroachment, and other fires to enhance ecological services or biodiversity. By itself fire cannot wipe away the brush. Nor can burning hold a site as though it were a mechanical mower. Fire will work best interacting with other agents—a cocktail approach that suits the plateau's style.

The opportunity, which is also a necessity, is to align burning with assorted biological controls, notably the region's traditional complement of browsers and grazers. A dandy research program has identified those breeds of goat that can browse young juniper sprouts at a time before terpenes render the buds toxic; the station is selecting and breeding the hardiest to help contain regeneration. While some seeding cedar can be killed outright, sprouting species send new shoots out quickly. The goats can hammer them and save grass for other stock. The gist of the project pivots around ecological reconstruction, not simply brush removal. Not least the goats can be sold. Brush cutters can't.

Here is the second fulcrum: the need for economic leverage. Barring government subsidies, which are becoming rarer, the project must pay for itself. Arguments about ecological integrity will have to be restated in terms of economic viability. What had supported the wool and mohair industries since 1954, the Wool Act, was phased out by new legislation passed in 1993. What were already marginal economics became even more tenuous. It makes little sense to burn out brush to promote grass or to use goats like an ecological atlatl if there is no market for sheep and goats, or if the economics favors conversion to recreational ranchettes rather than to pasture. (Further legislation in the 2002 farm bills helped stabilize the scene.) There must be a way other than government edict and state sponsorship to make private land serve a public good.

An Edwards solution is to toss up a salad of incomes. Improved pasture will, over a few years, yield better meat and fiber. Some brush may help by encouraging whitetail deer and turkey, which can be hunted. (The Sonora Station reaps 40 to 60 percent of its revenue from hunting leases.) Others may look to gas drilling. The problem is often the transition: to burn, the rancher must store up grass, or in other words, forego grazing. That loses revenue in the short term, and over the long term it means always fallowing a portion of the range so that burns can continue. Over the long run the land is healthier and the stock better, but there is an immediate cost, which may take years to recover unless some other revenue appears from the enhanced ecological goods and services.

The third reality is that both land and economics are changing. Practitioners may speak loosely about restoring the environment, but in truth they are remaking it. Everyone is better for lifting the heavy and oft-cruel grazing that had trampled the land under hoof. But while this has led to a more resilient herding in some places, in others it has meant the elimination of stock raising altogether. New owners may be absentee—investors, sometime visitors or future retirees, recreational hunters or birdwatchers. The state may buy some land for parks. What for over a century had been a more or less uniform, if harsh, ranching landscape, all of it subject to the same issues of brush encroachment, was fragmenting.

Typically environmentalists view fragmentation as a threat to habitat. In Texas it is often a sign of ecological recovery. Grasses revive, woods thicken, the landscape splinters into slivers of floral and faunal diversity. For all their ills the old practices had kept wildfire on a short leash because it had little more than scraps on which to feed. Only extreme drought and winds could push flames through the ungrassed brush. But the emerging landscapes of the Edwards have ample fuel, often where it will do the most damage to subdivisions and trophy homes. Like the montane forests of the West, fire exclusion had shifted the regime from surface fires to stand-consuming wildfires, and with juniper they can morph from wind-driven stampedes into plume-dominated conflagrations. If the slow combustion of livestock doesn't consume those fuels, the fast combustion of flames will. Given a chance, Texas grasses can rebound. So can Texas fire.

Ecology, economics, private land ownership, and fire all argue for a working landscape that must somehow pay for itself. Probably traditional commodities won't earn enough. So just as Texas has long externalized its costs, it may have to continue relying on an external income. The obvious source is the reservoir of gas underlying the plateau. The drill derrick is doing for recent generations what the windmill did for older ones. It may help get over the transition period.

Whether by intent or default, however, the land needs to be worked to prepare it for the right kind of fire and defend it against the wrong kind. Untreated patches sit on the land like oily rags in a corner, or vacant buildings into a derelict city block, ready to burn. They are a private choice that can become a public problem. Effective fire protection means fuel management, which sounds like someone telling a landowner what he can and can't do on his property. In Texas those are fighting words.

The real question, however, may be where that firefight takes place. If the Edwards Plateau follows the evolving Texas model the battle will pit new Texas against wildfire. Institutions like the TFS will help harden assets at risk and strengthen the community's ability to resist. If the Edwards exception continues to mature, however, a strategy of resistance may segue into one of resilience. It can substitute controlled fire for wild, and by making ranches better able to receive the right kind of fire or fire-browsing mix, the community overall will be protected. Neighbor will help neighbor not just by rallying when smoke rises on the horizon, but by having lands that cannot support the kind of fire that threatens one another. The more neighbors join, the more secure each member is. It's the fire equivalent of herd immunity.

Particularly within the Texas context, that process is hard to start. Just as prescribed fire needs banked grass, fodder taken out of production so it can feed flames, so institutions need something extra—some stored political will, a back 40 acres of social space, a bit of intellectual capital—to jolt it out of the daily scramble. The EPPBA has provided that institutional surplus, and Butch Taylor, the spark to set it ablaze.

COMBUSTION SLOW, COMBUSTION FAST, COMBUSTION DEEP (INDUSTRIAL TEXAS)

There is a scene in *Giant* in which Leslie Benedict decides to take a walk. Her new husband, Bick, tells her no one (or at least no one of any respectability) does that in Texas. They ride. The central conflict in the novel pits new wealth from oil against old wealth from ranching, but the disdain for walking endures through both. Today, Texas is a car culture (okay, a truck culture). You overcome distance by riding. There are few places to walk; the culture remains hostile to pedestrians. The man in a pickup has replaced the man on a horse as cavalier.

That segue is not a bad description of how fossil fuel has affected Texas and the Texas persuasion. It reinforced an existing matrix. It acted on already established habits of agriculture, frontier settlement, and politics. It flowed through old arroyos. The wealth from surface lands, in the form of livestock, was replaced or overlain by wealth from subsurface lands, as oil and gas; the landscapes just belong in the Paleozoic, and have been treated with similar roughness, and a tendency to substitute size for husbandry. The big man became the holder of subsurface or offshore leases rather than Mexican land grants or cessions to Texas empresarios. Pump jacks connect an ecology of industrial combustion to the old one of flaming grasslands like pushpins might fix a mylar overlay on a map. Pipelines bind like baling wire.

The old Texas of settlement times had a fire economy that pitted fast and slow combustion against each other. The land would burn, but livestock and fire competed for the grass, which would either slow-burn from metabolizing herds or fast-burn from open flames. The livestock mostly won, and this created an unbalanced fire economy that left an oft-degraded landscape that had too much grazing and not enough burning. The new Texas that resprouted from an economy based on fossil biomass and services added a third combustion into the mix, this one powered along a deep axis that runs through geologic time.

Today's Texas is a fire triangle of fast, slow, and deep combustion. Its lines of fire are its roads; its fields of fire are its power plants, petroleum distilleries, and electricity-guzzling malls and Astrodomes. In place of cattle, sheep, goats, and horses, it has Ford F-250s, lawnmowers, tractors, and 18-wheelers. Its fires feed on forage distilled from the ancient Earth and spew effluent destined for the planet's future. These energy flows bind and shape the Texas economy, which is to say, its economy of nature as well. They affect almost every attribute of the Texas environment.

Industrial fire has its own ecology. Fossil fuel combustion does not sweep over the land annually or triennially, but hourly, without regard to drought or deluge, year after year. It knows no burning period: it burns continually. It displays no seasonality: it combusts constantly. But its deep significance for fire history is that, so far, it has competed with the other forms of combustion and sought to substitute for them. Cattle or alfalfa can be shipped by rail or truck. Airplanes and tractors can spray herbicides and pesticides to replace the purgative effects of flame. Where fire had to fertilize by pyrolytic decomposition, artificial fertilizers can do it by surface application. Internal combustion machines spread herbicide and nitrogenous fertilizer. They cycle flora and fauna. They fight free-burning flame.

The advantages to daily life are many. Farm wives no longer have to cook or wash over stoves, sweltering through the summer. Fresh foods are available year-round in air-conditioned stores. Errant sparks are less likely to threaten winter range. Transmuted into pistons and electricity industrial combustion has freed up time from daily life. If the power it produces is replaced, the new sources will likely be renewables such as solar or wind, not a return to open fire. Once raised in a fossil-fuel world few people choose to renounce its benefits and return to smoky kitchens, the drudgery of endless wood gathering, and the dangers of escaped sparks.

But that is the world of quotidian life and work in fab plants, suburban subdivisions, and office parks. For the Texas landscape the choice is reversed. Much of it needs fire, and the task is to find ways to have industrial combustion free up space for flame as it has liberated time from domestic routine. On an individual ranch, supplemental oil and gas can create surplus money that can allow grass to be banked for burning (although it can equally be used to keep cattle on the land during droughts, which helps the economics of ranching but degrades the grasses). What has yet to happen is a collective analogue that will work over larger landscapes. Across the United States the existence of public lands allows for that surplus of space and money. In Texas there is no comparable mechanism. Rather the tendency is to bolster the existing relationships, not recharter their fundamentals. Hardscrabble ranches may become vineyards, and rotating pastures may yield to strip malls, but there is little move to transfer land out of active production and into nature's economy.

Whether by deliberation or inattention new Texas is freeing up surface fuels, which is to say, it is creating conditions for free-burning fire. The ideal strategy would seem to burn that "surplus" off under controlled circumstances, to treat those liberated fuels as fallow in nature's economy and periodically burn them off under controlled conditions to jolt the land back to life. (There is no reluctance to flare off natural gas from wells.) Those restored fires would go a long way to repairing the degradation caused by fire's absence, as well as dampen the prospects for conflagrations.

In brief Texas needs to rebalance its emerging fire triangle. A three-legged stool is stable. A stool with one leg far larger than the others will topple.

In the new Texas economy of fire two expressions seem particularly revealing. Both convey the relationship between the three combustions, and both illustrate the way Texas joins with the rest of the nation.

One is the state's electricity network, which along with its asphalt roads and flow of vehicles aptly inscribes a matrix of deep combustion. Alone among states or even regions Texas has its own grid, which connects to the eastern and western grids that collectively comprise the United States network. The Texas system is not complete in that the Panhandle and Houston stand outside it, but its coverage is remarkably coexistent with the state's borders. It's a suitable symbol of the state's blend of independence and connectivity. And it describes equally well the state's relationship to the rest of the country with regard to wildfire. Texas has its own intrastate institutions, and only links to the nation during overloads and emergencies.

The second expression pertains to the peculiar linkage between industrial combustion and wildfire. Increasingly the realm of deep combustion is reducing the domain of slow combustion, with the unexpected outcome that fast combustion is reclaiming some of its former dimensions. The perfect symbol of this unwanted linkage is the power line that, in high winds, fails and kindles a fire in the altered landscape beneath it, a landscape increasingly lumpy with surface fuels and houses. That the brush is storing the excess carbon liberated from burning oil and gas, and that the droughts and winds are tied to a climate that is being unhinged by overburning petroleum—the combustion equivalent of overgrazing the once-open range—may go beyond irony. It is enough to note that the most damaging conflagrations of recent years, notably the Bastrop complex, have resulted from the violent arcing of electric wire to grass and brush amid a landscape remade by a fossil-fuel civilization.

This is clearly one way to join the two realms of combustion. But there must be a better one.

RED NORTHER: BASTROP BURNS

On Labor Day weekend, 2011, a centennial drought, high temperatures, fuels heaped like back-40 woodlots or the biotic equivalent of auto junkyards, and vicious northern winds conspired to drive flame through the Lost Pines region east of Austin in what might aptly be characterized as a red norther. What became known as the Bastrop complex climaxed the worst fire season in Texas history.[23]

The basic facts seem to speak for themselves.

On September 4, the Sunday of Labor Day weekend, dry winds from Tropical Storm Lee, which made landfall to the east, toppled a dead loblolly onto a power line and kindled a fire. The call to the Bastrop VFD came in at 2:20 p.m.; the first engine arrived five minutes later and instantly requested backup from the TFS. The emergency operations center informed them that there was little to send. Other fires—the TFS was already reinforcing nine in central Texas—had stripped the system of any surplus resources. After eight minutes, with the fire roaring out of control, the crews pulled back. A half hour later another falling tree started a second fire four miles away. When the flames crossed Texas Highway 21, the county officially declared the fire a disaster. Neighborhood evacuations commenced. The fire blasted into Bastrop State Park. It leaped highways, the Colorado River, ranches and developments, and Highway 71.

The flames burned beyond anything anyone on the scene had witnessed before. At times they cascaded up like a colossal burnt offering, and at other times they seemed, even to fire specialists, like a mutant phenomenon that drove them to search for analogies beyond those encoded in fire behavior models. The spectacle seemed to some less like a wind-driven fire than a pyroclastic debris flow, a turbid current of ash, embers, and gases that outran radiant heat and washed through fuels. It burned as straight as the blast of a blowtorch, and reports pointed to horizontal roll vortices that appeared to hold the pyroclasts into a ground-hugging plume much as the magnetic field of a torus contains a plume of plasma, and when the field faltered, its confined contents exploded outward, alternately constricting and collapsing the flow from hydraulic jet to consuming splash. By 5:00 p.m. the two fires, rushing south, merged. A third fire began through the same mechanism and later merged with the others. By nightfall the Bastrop complex had burned an oblong swath some 6 by 14 miles. The next day another fire broke out; the winds veered and the fire front moved southwest; the flames chewed on. Then the winds died. The counterforce of engines, planes, and crews began to crest. The fire quieted, and on Wednesday it ceased to spread. By then FEMA fire assistance grants were activated. Full containment took another 23 days. Not until October 10 was the Bastrop complex officially declared out.

The reckoning included 34,356 acres, 1,645 homes, and 2 lives. The blowup climaxed a dismal roster of Texas records—the second-longest fire season, the hottest summer, the driest year since 1895, the most catastrophic losses. Almost four million acres burned and over 5,600 structures were lost. Bastrop State Park, the crown jewel of the Texas system, burned all but 250 acres. The mutual-aid system that had evolved since 1996 broke down, strained far beyond its worst-case scenario. The plume, like a tornado on its side, rushed south within eyesight of the state capitol and what was widely touted as a site of high-tech synergy. Among the most elemental expressions of nature's economy had met the next new thing in the nation's, and fire had won.

Yet the horrors could have been far worse. The fire was off the scale: no system could have expected or prepared for it. A decade and a half of institutional improvements had readied a far wiser and safer response, which had coped as well as possible. Officials called the evacuations right. VFDs that might have found themselves overrun had shrewdly pulled back and regrouped. Cleanup and rehabilitation proceeded quickly and knowledgeably. What it could not have stopped, the system accepted and then rapidly picked up afterward.

The looming question, in other words, was whether the blowup was an outlier or an augury. The possible interpretations are many.

The Bastrop complex was a fire in the same sense a *Tyrannosaurus rex* is a lizard. What stunned was not just its size in a region not known for big fires but its ferocity and the unspeakable damages it inflicted. What puzzled, and frightened, observers was what the outbreak meant for the future, the uncertain prospect of whether Bastrop was a freak of nature or the future of fire in Texas. With symbolic aplomb it sat on the border between south and west—and perhaps on the border of past and future. Was it an isolated outpost that defied defense (an ecological equivalent of the Alamo), or a corridor, a contemporary camino real, through which modern settlement would pass? Was the burn a kind of Dust Bowl moment, a horrific but focused flare-up of climate? Or was it the equivalent of a lurch in a receding ice mass that signals a fundamental realignment of geography? Either or both constructions are plausible. Seasonally hot, episodically droughty, always laden with grasses if not woods, ever windy, it would not take much to topple the landscape into a new and catastrophic regime. For now the questions are metaphysical, and inquirers are likely to search for metaphors by which to extract meaning out of the ash.

Those inclined to draft symbols from nature could point to the Lost Pines. Roughly 170,000 acres of loblolly pine, they lie apart from the main swath of southern pineries like an offshore island. Genetics suggests they are a Pleistocene relic, a large patch that somehow managed to survive and over time has undergone its own evolutionary selection to sharpen adaptation to its peculiar, slightly more arid setting. In this view the Lost Pines exist on a rim of extinction, like condors or sequoias; the cost of sustaining them will always be high and the odds long. Fires may drive them over the brink. They are too small to survive against enormous changes in climate and human settlement. Other fires may yet follow, but the biotic isle is finite, and something new will emerge out of the reformed landscape. Big Bastrop burns are a signature of the Anthropocene. The pines will be lost.

Those inclined to extract symbols out of human contexts will attend to the dynamics of urbanization and habitat fragmentation. Bastrop County is where a pattern of suburban and exurban sprawl, typical of the wealthier South, was overlaid with the droughts and fires typical of the West. The Lost Pines were less a coherent biogeographic entity than a pastiche of private lands devoted to ranchettes, hunting blinds, trailer parks, strip malls, recreational cabins, and a couple of small state parks, much of it so choked with pine, oak, and cedar it could stop an Abrams tank. Bastrop was where the fire patterns typical of the West met the pyrogeography of a newly drought-blasted South. When the modern era of Texas WUI began at Cross Plains, the Texas Forest Service likened the outbreak to the 1991 Tunnel fire in Oakland, California, and treated it as an annunciatory event. The Cross Plains conflagration consumed 117 structures; the Bastrop complex, 1,645. For Texas the Bastrop burn fundamentally changed the calculus of fire management. In this scenario the pattern will repeat until landowners, or an outraged public, intervene to reconstruct how they live on the land.

A third option is to draw symbols out of the past. Bastrop is in cameo where South met West, and thus it stands for how Texas took a southern model of settlement and adapted it to the progressively arid landscapes extending westward. It was where two patterns of fire, that of woodland and that of prairie, collided. It has remained a place where contemporary southern settlement patterns sprawling through the woods have thrust into a drying environment, and thus where the urban-rural and the feral-wild meet. Among Texas's many fire disasters in 2011 those where east and west met were the worst; and since the tension was greatest at Bastrop, its conflagration exceeded the others. What makes this symbolism into a cipher, however, is that we might be watching history's flow reversed. Instead of east to west it may be that extreme fire is moving from west to east, along with the methods for containing it. In this imagining Bastrop is not an outlier but a portal through which the past will flow into the future.

Whatever the outcome a final reckoning will depend on where Texas fits into the national fire narrative. It may be that Texas is where, for America's fires, the West meets the South.

EPILOGUE: TEXAS BETWEEN TWO FIRES

What I am trying to say is that there is no physical or geographical unity in Texas. Its unity lies in the mind.

—JOHN STEINBECK, *TRAVELS WITH CHARLEY IN SEARCH OF AMERICA* (1962)[24]

My uncles sat on the barn and watched the last trail herds moving north—I sat on the self-same barn and saw only a few oil-field pickups and a couple of dairy trucks go by. . . . And yet, that first life has not quite died in me—not quite.

—LARRY MCMURTRY, *IN A NARROW GRAVE: ESSAYS ON TEXAS* (1968)[25]

Critics and partisans both have long regarded Texas as a state of mind. Its physical geography is a setting, not an informing principle. The defining feature of its landscape is simply its size. Instead what fuses Texas society is a story. How Texas fire connects with the rest of America will require linking stories as much as transferring air tankers, hotshots, and incident management teams.

Once planted the Texas narrative has proved as tenacious as its brush. The state's writers divide roughly along the eras of old and new Texas. Old Texas is a place of folklore and nonfiction. New Texas is the scene of popular culture and fiction, although both seem unable (and unwilling) to drop the reins to the past. The state's most renowned writer, Larry McMurtry, has spent a career trying to kill off old Texas—literally, in the case of his first novel, *Horseman, Pass By*. Despite successes in writing (and scriptwriting) about modern life, he admitted early "that the place where all my stories start is the heart faced suddenly with the loss of its country, its customary and legendary range."[26]

The hugely successful *Lonesome Dove* series returned to that country of the heart, but with a gruesome, and ironic, retelling that climaxes with Clara berating ex-Ranger Woodrow Call for letting the past override the present. Captain Call is honoring the request of his compañero, Augustus McCrae, to be buried in Texas, and is returning a maimed corpse from Montana while leaving his unacknowledged son, Newt, behind to manage without him. Clara denounces that choice. Newt is the future,

Gus the past. But Call almost mindlessly persists and the horrors trail along with him. And so, in a sense, has audience desire for the old narrative triumphed over the new, despite an enormous influx of newcomers, because it is coded both overtly and covertly into Texans' way of life. No matter how often it is top killed, the story resprouts.[27]

That same troubled continuity affects the narrative of Texas fire. The old legends are hard to grub out. A new narrative that can usefully link with the evolving national scene will likely have to be grafted onto that enduring rootstock.

The old Texas that crushed fire out of its landscape still claims much of the state. If, as the saying went, Texas was hard on horses and women, it was also hard on land. The fires that the Tejanos once feared, they grazed and beat out of the landscape. Yet their grasses were just as tough. They survived, although they steadily degraded as owners each year spent a bit more out of their ecological capital. As grazing's heavy grip loosened, the land rebounded. The outcome was not always as desired and rarely as planned, but it came in ways that recouped a few of the losses and created some space for fire.

The new Texas that began swapping out the landscape tiles of the old belongs squarely in the realm of industrial combustion. It is refashioning the homogenized scene into a lumpy one. Of the disheveled landscapes that are emerging some are intentional, some have happened inadvertently, and some follow from simple inattention. A fraction of the land has moved into the public sector, though not much. More is passing into fallow from freed pasture or farmlands formerly watered by now-drained underground reservoirs, or are devolving into absentee ownership, or to holders who value hunting more than herding, or those for whom a ranchette is a trophy of having arrived in Texas society. Whatever the cause, once removed from intense production, the land is regrowing and is available for other purposes.

It is unlikely that it will become wildland—that concept doesn't register in the Texas psyche. Open range isn't wilderness; the unoccupied land is land available for seizing. Nor will it be public land, a commons for a greater good. It will be worked land under private ownership or perhaps public-funded easements. What isn't worked will become feral, as will its fires. The trick will be to redefine the working landscape into one promoting ecological goods and services, and to find a way to mark them to market, and to somehow craft a cooperative spirit that can find a common cause if not a commonwealth.

In the meantime Texas will experience further outbreaks of wildfire, and its instinct will be to attack them, to send in a company of Rangers and kill the bad guys, which will make the problem disappear. That won't solve the fundamentals, but it brings a sense of immediate security. Likely the state will not be able to do even that by itself. The future character of Texas fire will depend on how Texas connects with the rest of the country.

The problem with exceptionalism is that it doesn't transfer—that's what makes something exceptional. The question with Texas exceptionalism and fire is determining how in fact the Texas scene and strategy might relate to America overall.

Those linkages will include narrative. It seems nearly impossible for Texas to shake off its own story of itself, despite its recent wave of immigration, much from Mexico, and its status as America's first minority-majority state. The story is too deeply rooted: even when it is not visible it casts shadows. But that story does not speak widely to others. In truth there are many Wests. All the others have their own relationship to land, and to fire. The Northern Rockies have a western narrative in which Americans and Canadians competed over the fur trade, and are more inclined to celebrate Lewis and Clark than Goodnight and Loving. The California West looks nothing like Texas's: it barely looks western except in its scenery. Oregon's pioneers, trekking by wagon along the Platte, bear little resemblance to drovers crossing the Brazos. By comparison to Texas's origin story, Utah's (a wholly western one) might as well be sited on an outer moon of Jupiter. The Texas cowboy got into Hollywood. The Texas fire officer may find it hard to cross the Red River.

Nor is the Texas master narrative one sympathetic to environmentalism. Rather like Brazil, Texas has defined itself by its people and their story, and its geographic setting has mattered mostly because it was too large to absorb. Its sacred places are historic sites, not natural marvels. Its symbols are the silhouette of the state, a Lone Star, and the Alamo, not something from its natural heritage. Elsewhere public lands act as a fulcrum to force environmental concerns into public consciousness. Texas avoided such conflicts and has had to wangle among landowners and an ideology of private property instead. In place of state-enforced environmentalism (much less notions of deep ecology), Texas opted for beautification and screening off junkyards from highways. It's okay to mess with Texas by salt flushing a drill site. But toss a hamburger wrapper onto roadside bluebells and you may face a $2,000 fine.

The style of Texas politics doesn't help. Texas has been adept at extracting money from the federal treasury, often funneled through state institutions. It has tended to treat the United States much as France does the European Union, as a means to protect Texas interests and project Texas values. Both seem to exaggerate out of an alloy of chauvinism and insecurity about their place in a larger world. Much as Texans resent the imposition of values from elsewhere that sit awkwardly in the Texas setting, so Texan matters can ill suit the world beyond. The two native-born Texans who have been president in the postwar era each led the country into trumped-up foreign wars (which they lost), trashed the economy, and polarized the commonwealth. The Texas story, the Texas landscape, the Texas persuasion—none seems to travel easily outside Texas.

The record suggests that any exchange of fire with the rest of the country will be awkward, and so it proved in 2011.

As the fires kept coming, as one start after another leaped over roads and blew past VFD engines, as the TFS exhausted what it could move around the state under the Texas Intrastate Fire Mutual Aid System, Texas had to look farther afield. It had done so, grudgingly, for almost 20 years. Now the force arrayed against it was overwhelming. It ordered national resources and incident management teams.

Because the request passed through the Texas Interagency Fire Center, the state essentially looked to its origins and ordered a team from the southern region. The Southern Type I IMT blue team, however, went to the trans-Pecos where it found itself in new country. It had to cope with fires that blistered past everything it threw at them. It had to organize VFDs and ranchers who had their own ways of fighting fire and understandings about who was in charge and who might legally join the firefight. They had to deal with county judges and commissioners who insisted that they, not incident commanders, held final authority and would decide, not delegate, what would be done, when, and how. Meanwhile the winds blew, the evenings stayed dry, and even the Davis Mountains presented no more barrier than a patch of tobosa grass. Historic Fort Davis, established in 1854 to defend against Apache and Comanche raiders, had never faced such an assault.

A million acres in flame, with townsfolk in wholesale evacuation and national historic sites and state parks under attack, was not the time and place to negotiate a new compact between Texas and the United States fire establishment. When their tour ended, the blue team cycled out, and the southern area red team replaced it. The winds calmed, and tempers with them. When the season ended, all parties sought to recharter the system, although each interpreted what that meant differently. The locals wanted more money and more say. The feds demanded adherence to national standards even if local circumstances might argue for accommodations. As usual the state found itself in the middle. The National Cohesive Strategy, then working through Phase II, has proposed a matrix for just such discussions. As the adage goes, no battle plan survives contact with the enemy, but the revisions catalyzed by 2011 will almost certainly make the next engagement less contentious.

What no one provided, however, was what everyone needed: a narrative that can connect the Texas experience with that of the nation. Yet one may exist, latent in the state's founding story: Texas is where the South passed through to the West. The Texas revolution was the source for the phrase "manifest destiny." And the dryline between east and west Texas was where a hybrid ranching culture emerged that carried destiny westward. Those two events provided the founding stories of the state. One might, in a fire-history paraphrase of McMurtry, sit on a windmill and watch ranch hands with burlap sacks and beef drags beat fire out of the land.[28]

What may bring Texas back into the national scene is that, with respect to fire, the historic flow seems to be reversing. If, as the prophets of global change fear, its landscapes warm, its woods dry, and its fallow farmlands overgrow with brush and houses, the eastern United

States will burn and may well become the future national frontier for fire. The model it needs to cope with that specter will come from the West, but it may well pass through Texas along the way. Central Texas could become the place where a fire-management system adapted to the arid West retools to cope with a more humid, more populated, and vastly more complicated political landscape eastward. One might, akin to the younger McMurtry, sit on a fire tower and watch the passage of hotshots and helitankers.

It's an arresting thought: Texas as the portal through which the West passes through to the South. Of course it may not happen, may never be needed, and in any event won't be easy. It would demand a process of mutual mythmaking as much as negotiations over delegation agreements. But the prospect suggests a way Texas might comfortably redefine its historical role and why the Texas experience might matter to the rest of the country apart from drains on the national treasury and emergency claims to suppression resources.

A long trail—but one that leads through Texas in a way that allows Texas exceptionalism to reassert itself. Texans wouldn't want it any other way.

OUTLIERS

PEOPLE OF THE PRAIRIE, PEOPLE OF THE FIRE

TWICE OVER THE PAST 20,000 years the Illinois landscape has been destroyed and rebuilt. In the first age the agent of change was ice, mounded into sheets and leveraged outward through a suite of periglacial processes from katabatic winds to ice dam–breaching torrents. The ice obliterated everything, leaving as its legacy a geomorphic matrix of dunes, swales, moraines, loess, great lakes, and landscape-dissecting streams. For the second, the agent was iron, forged into plows and then into rails. Coal replaced climate as a motive force, and people pushed aside the planetary rhythms of Milankovitch cycles and cosmogenic carbon cycles as a prime mover. They left behind a surveyed landscape of squared townships.[1]

The first event worked through a geologic matrix; the second, a biological one; and they were equally thorough. All the state went under ice at least once; the last outpouring, the Wisconsin glaciation, pushed south from Lake Michigan and covered perhaps a third of Illinois. The frontier of agricultural conversion put nearly all of the state to the plow, or, where rocky moraines prevented it, to the hoofs of livestock. When it ended, only one-tenth of 1 percent of the precontact landscape remained more or less intact. Less than one acre out of a thousand held its founding character, and that acre was itself minced into a thousand, scattered pieces.

In both Ice Age and Iron Age, however, life revived after extinction with fire as an informing presence—fire in the hands of people. The biological recolonization of the landscape after the ice had fire in its mix and expressed itself as oak savannas, tallgrass prairies, and grassy wetlands, stirred by routine burning. Fire was a universal catalyst; in particular, prairie and fire became ecological symbionts. The reconstruction of the second landscape has relied on industrial combustion, fueled by the fossil fallow of biomass.

But those intent on sparing, or actively restoring, the former landscape must appeal to open burning. A fire sublimated through a tractor does not yield the same effects as one let loose to free-burn through big bluestem. The regeneration of such settings is troubling—unstable and scattered, an inchoate genesis still in the making. Its reliance on fire is both essential and challenged.

The indigenes at the time of European contact, the Potawatomi, were known variously as the people of the place of the fire, or the keepers of the fire, because they maintained the great council fire around which the regional confederation of tribes gathered. But that fire did not stay within the council circle. It spread throughout the landscape, a constant among the diversity of grasses, trees, shrubs, ungulates, small mammals, birds, and insects that congregated around the informing prairie. In time the Potawatomi became known equally as the people of the prairie since the one meant the other. Remove fire and the prairie disappeared. Remove prairie and free-ranging fire lost its habitat. Remove the keepers of the fire and both prairie and fire vanish into overgrown scrub, weedy lots, or feral flame.

Restoration is a slippery concept. In some places it means mostly finding ways to preserve and enhance relicts that have survived the battering. In other places it means an outright regeneration, or a reconversion of farmland to prairie. But at its core it involves sparing the pieces and saving the processes that connect them. In Illinois, once the prairie state, now a factory farm, prescribed burning is what connects those pieces, and prescribed burners are the agents that join them.

KANKAKEE

The unity of the Kankakee sands region lies in one of those convulsive geologic aftershocks of glaciation. As the Wisconsin ice receded, it melted, and the meltwaters ponded behind berms of moraine or lobes of adjacent ice sheets. Eventually those dams themselves melted or were breached, and the impounded waters drained out. This often happened catastrophically in the form of floods or, in local parlance, torrents. At Kankakee the outrush left a scoured landscape of sand dunes and wet swales and incised streams. It became an archipelago of soils and landforms whose connection looks back to events 17,000 years in the past.

Each site took on additional characteristics as the result of its recolonization and, during the second—the settlement—torrent, the ways in which it was farmed, drained, grazed, or subdivided. Historically the lowlands were marshy and grassy, and the uplands more forested. But extensive draining converted the swales into corn fields, while routine burning kept the uplands in a woody savanna—the largest remnant of extant oak savanna anywhere. Critically, while grazed, the uplands were not plowed: their soil structure remained intact. And, exceptionally, they continued to burn.

The great northward migration of African Americans had an echo in a secondary outflow from Chicago, a city some found too alien and job poor, into subdivided lots around Pembroke. There they settled down amid old habits, including casual fire, and an absence of government services, not least fire protection. The lack of trash collection, in particular, meant they burned refuse, and these fires frequently escaped to kindle the countryside. The surrounding sand ridges burned roughly every 1.5 to 2 years. An area of extreme economic poverty became, paradoxically, a place of exceptional biotic wealth.

Today that miscellany of missed places constitutes an atoll of natural areas, some 32,000 acres in all, allocated among 33 designated sites, hopefully labeled the Greater Kankakee Sands Ecosystem. The archipelago includes Goose Lake Prairie State Natural Area, Des Plaines Conservation Area, Midewin National Tallgrass Prairie, Wilmington Shrub Prairie Nature Preserve, Laughton Preserve, Mazonia-Braidwood State Fish and Wildlife Area, Iroquois Woods Nature Preserve, Mskoda

Land and Water Reserve, Sweet Fern Savanna Land and Water Reserve, Kankakee Sands Restoration Project, and Willow Slough Fish and Wildlife Area, and with those sites a roll call of Illinois conservation organizations that ranges from national agencies to state and county bureaus to NGOs: the U.S. Forest Service, the Fish and Wildlife Service, the Illinois Department of Natural Resources, the Nature Conservancy.

In all this—remnants scattered like lithic flakes, restoration projects sprouting from corn stubble, a variety of institutions as diverse and dispersed as their biotic relicts—Kankakee is a cameo of the Illinois conundrum. No single site, institution, or vision contains it all or organizes the pieces. There is no commanding height—not topographically, not institutionally, not intellectually. A federal presence is muted, quarantined on checkerboard hills in the far south. There is no domineering private landowner—no Weyerhaeuser, no Ted Turner—to deform the space-time of land use. There is no counterforce to challenge the industrial plow. What the pieces and players share is a variously defined commitment to nature protection. They are, like the Potawatomi, peoples of the prairie, which means they are also peoples of fire.

They differ in goals. Some believe that the task demands a way to connect the fragments into a whole, at least conceptually; they seek out corridors to join the parts, or ideas to help identify which pieces should be protected in what order. Others believe that salvation depends on size. Unless the protected areas are large, unless they contain within themselves all the required parts, the whole cannot hope to survive against fragmenting forces of regional or continental scope, not to mention globalization. Yet the practical scale of either strategy is so small that the atolls they oversee may both be drowned in the rising sea of a modern economy. Chicago adds more rambling exurbs yearly than the state does protected preserves. Farmland converts to city, not nature.

Each site resembles a miniature, the ecological equivalent of a ship in a bottle. Its minuscule scale allows for some processes to persist and for the abolition of known destructive practices. But they struggle to become a whole; the separate parts cannot absorb the roaming elk and bison (and successor cattle), or their predators, that helped define the historic scene. Their collective fauna is one that travels by air, and that is also tiny; the faunal diversity consists of birds and especially invertebrates.

This can cause troubles, however, because insects can be highly specific in their preferred habitats, and they can attract partisans that consider butterflies and leafcutters in old-growth prairie as the counterpart to spotted owls in the old-growth forests of the Pacific Northwest. The species triumphs over the habitat. In order to accommodate, even small plots might have to partition into micromanaged patches. A landscape that boasts white deer and wolves must shrivel to one for beetles and the regal fritillary. This matters because some management practices cannot be indefinitely shrunk, any more than Newtonian physics can scale evenly from quasar to atom. A butterfly and a bison demand different minimums of place.

So, too, does fire. In a miniature landscape it acts more like a blowtorch than a free-ranging wind. It behaves like an implement of horticulture, a clipper or hoe, no longer feeding itself as it propagates but consuming what it is served. The patches resemble cages in an open-air zoo or, to mix in a more benign metaphor, like rooms in a hospice. The ecology of a candle bears little kinship with that of a prairie aflame. No one knows the scaling laws for fire ecology that might join the nanoniches of a prairie refugia to a boreal crown fire. They only know they must have fire.

This, however, is the second element the system shares: a commitment to burning. Fire does nothing here it does not do elsewhere. It just seems more prominent because it is indispensable and the small scale of the operation makes it undeniable.

The remnants survived because they were burned. If flame leaves, woody plants will quickly swarm over and smother prairie and savanna. A handful of years is sufficient to let invasive shrubs and trees establish themselves to the point that fire alone can no longer knock them back. Like a boa constrictor steadily tightening its coil whenever its victim breathes out, the woods crush the grasses and forbs when the pause between fires lasts too long.

Questions of scale do not, however, abolish all principles of fire ecology, and one is that organisms do not adapt to fire in the abstract but to a fire's regime. You can lose a site as surely by burning badly as by not burning at all. The bouquet of sites around Kankakee argues for a bouquet of burns; and in the absence of particulars, a useful rule of thumb is a three-year rotation, which approximates the core cycle of postfire recovery and, at Kankakee, will accommodate almost all species if a site is large enough relative to the organism's demands. Still the threat of too little fire probably trumps the threat of too much. The premier relicts like those around Pembroke burned almost annually, or no longer than biennially; they burned as frequently, that is, as fuel existed to carry the flames.

Another principle is that fire is *bio*technology. Its flames do more than act as a fiery brush cutter; beyond merely mowing and mechanically rearranging, they transmute. They chemically change the biomass they consume, as grazers do. Nothing else provides their range of ecological services. Moreover, add to the roster of precepts that fire is an interactive technology as well. Fire's effects rarely result from fire alone but from the cascade of interplays it sets in motion, among them grazing and browsing, both gone from the Kankakee complex except at the level of insects. That is why mowing or planting or shunting fuels around cannot substitute. You can't replace clipping for burning as you can an electric bulb for a candle.

All this argues for prudence, which here means perpetuating the regime that kept the sites intact. To the uninitiated the precautionary principle might seem to argue for the opposite: they might invoke it to halt burning until a full-spectrum ecology can be worked out. But of course a complete ecology will never be known, and while species-partisan Neros and academic researchers fiddle, Rome won't burn, with disastrous results.

From the grand perspective of fire's recession and recovery across the American continent, the kindlings around Kankakee seem quaint, a daub of mom-and-pop stores amid an economy of big-box retailers and multinationals. This is boutique burning, almost a farmer's market of handcraft fires. California's Cedar fire burned an order of magnitude more acres in one savage surge than the Greater Kankakee has under its collective protection.

But such metrics miss the point. It is not the number of scorched acres but the richness of their impact that matters. Acre for acre, probably more comes out of the Illinois burning than from all the firing of southern loblolly pine plantations and of generic western wildlands. Here, fire is the critical catalyst, without which the land cannot be defibrillated back to integral prairie and healthy savanna. Fire alone can't make that restoration work, but nothing done without fire can succeed.

NACHUSA GRASSLANDS

The Kankakee complex is a strategy of small parts in search of a larger context. Its conceptual counterpart is a large preserve that might contain the varied habitats of interest within its own borders. In Illinois its prime expression is the Nachusa Grasslands managed by the Nature Conservancy. But Nachusa is becoming large (in relative terms), not because it has preserved an ancient landscape, but because it is rebuilding one.

Its core is a small ripple of rocky hills that escaped the plow and hence retained some native species. In 1986 TNC purchased 400 acres. But around that atoll lay a platted sea of row-cropped maize. Over the course of 20 years the preserve has expanded to 2,000 acres, all purchased from willing sellers at market prices. Those acres must be brought into the system through a laborious process of restoration. By 2009 some 93 projects, ranging from 2 to 60 acres each, had expanded the dominion of prairie. Half the labor has come from volunteers.

It isn't enough to leave the acquired land to nature's touch: it will grow weeds. Instead prairie must be cultivated with even more tending than commercial crops; and there is little natural about the boundaries, which must follow the square-surveyed townships of settlement and the economic rhythms of commercial agriculture. Until they have the wherewithal to begin the conversion, they lease out the land for corn. Meanwhile stewards gather seeds from existing prairie.

They begin actual restoration—or more accurately, a reconstruction—by harvesting the corn, burning the stubble, and sowing a heavy mix of native seed, as much as 55 pounds per acre of 150 to 200 species. The next year assorted plants will have rooted, along with a street gang of weeds. Some weeds are an immediate concern and are clipped and individually herbicided. Others will succumb as the indigenes thrive. What matters is removing the nasty species and stimulating the desired ones, especially the grasses like little bluestem, since they will carry fire over the plot, and fire is what sparks the system to life.

The overseers burn as often and as intensely as possible. Where desired plants flourish they may overseed with more, and where some seem lacking, they may try again and let nature determine the suitability of niches. Meanwhile they burn. By the third year of tedious culling a raw matrix of prairie grows on the site. When they determine the mix is more or less right managers can begin backing off annual burns and feel their way toward a suitable cycle. Fires spread across the surveyed borders and suture the larger quilt of patches together. In this way the restoration reverses the frontier inscribed under the parameters of the Northwest Ordinance of 1787; and with so much work done by volunteers the process resembles a kind of reverse homesteading.

The contrast is not only with Kankakee but with such long-standing prairie sites as Konza in the Flint Hills of Kansas, gazetted in 1982 as a Long-Term Ecological Reserve. Konza is the classic model of nature protection and its servant science: it was a preserved landscape (and hence "natural") that kept prairie continuously under a regimen of burning, and later grazing. Nachusa is a reconstructed landscape, resuscitated out of corn stubble and ragweed. As it challenges Konza in size, however, it may also challenge Konza's embodied conceptions of what constitutes prairie and what deserves sustained research and perhaps may come to merit LTER standing as well.

The fire story at Nachusa is simple enough to state. Fire initiates the conversion, and once it has worked that alchemy repeated burning perpetuates the revived biota. Restoring prairie has meant restoring fire: this much is unexceptional, however quirky the process might appear to deep ecologists intrinsically wary of Roundup and flame. Rather, Nachusa's natural character resides in its present expression, not its history—or as William James famously described pragmatism, "By their fruits ye shall know them, not by their roots."

Yet there is a second narrative of fire restoration at work as well, in which fire is returned not only to the land but to the hand. The reconstruction of Nachusa reinstates fire to ordinary people. The volunteers who do much of the hard work of gathering and disseminating seeds, clearing invasive shrubs and weeding new acres, also do the burning. As much as reinstating big bluestem and lady fern, Nachusa has returned the torch to folk practitioners, the kind of fire wielders who sustained the prairie peninsula through millennia. The people of the new prairie have become people of the new fire.

This is a story easily lost among the attention paid to the traditional big hitters of fire management, and it counters two trends. One is the grand narrative of earthly fire by which industrial combustion has replaced open burning through technological substitution and outright suppression. This is why there is no fire on the still-farmed lands around Nachusa, why cooling towers from a nuclear power plant loom over the northern horizon from the rebuilt barn that constitutes preserve headquarters, and why quads and tractors rather than draft animals fill the sheds. Nachusa is putting fire back on the land.

The other trend is the systematic stripping of fire from the hands of the folk. The simplistic yet orthodox narrative for justifying the restoration of fire on public wildlands is that nature had set fires and misguided public agencies extinguished them, and the outcome is the shamble of present-day fire regimes. Such a narrative implies that restoration means no longer suppressing nature's fires. It means that people have to quit interfering with nature's logic. Nature will then begin deleveraging the landscape into it proper state.

Yet the record for virtually every landscape is that people had set most of history's fires, and this leads to the conclusion that the missing fires—those that have disappeared over the past century—are the result of people no longer acting as we have acted throughout our existence as a species. Less and less burning got done because there were fewer and fewer burners to do it.

To be sure, not all of that erstwhile burning was prudent or systematic; some was abusive and promiscuous and not a little simple fire littering. But in shutting down, the excess fire became, in effect, a government monopoly, something so seemingly arcane and technical and intrinsically dangerous that ordinary citizens could not be trusted with its stewardship. In this narrative, restoration means getting people to burn again. What Nachusa adds is the return of the torch to private citizens, not solely to agents of government.

RECESSION AND RESTORATION

The ice age receded, across a span of 10,000 years, with a succession of geologic spasms like the Kankakee torrent. The recolonization of that evacuated landscape by life took several millennia, and after the climatic maximum of the Hypsithermal, humanity helped stabilize its dimensions and the resulting pastiche of prairie and savanna by regular firing.

The Iron Age ended with soils bleeding from a thousand thousand cuts and with a slow smothering beneath a blanket of domesticated flora. Its regeneration will take centuries, if not longer, quarter section by quarter section, township by township, and it will act out against a fast-morphing climate, likely the byproduct of an industrial burning run amok. But it will happen at the hands of a humanity wielding fire.

This is not the kind of creation story or heroic narrative that American environmentalism has traditionally thrived on. But it is what must happen if nature's economy is to continue to produce the goods and services we want. It's a story in which the Hippocratic injunction to first do no harm means you will harm if you don't first do. And it's a story in which the people who want prairie must also become a people who want fire.

WICHITA MOUNTAINS

ASK THE FIRE COGNOSCENTI about what places in the Great Plains have fire more or less right, and you will be pointed to the Flint Hills and the Wichita Mountains.[1]

They straddle the 98th meridian, one east and one west, and at first blush, it is their differences that strike the mind. The Flint Hills are a hump of cherty limestone; the Wichita Mountains are the gnarled granitic stump of an ancient mountain range, a Black Hills of the southern plains. The Flint Hills hold the greatest remnant of tallgrass prairie, grading southward into the Cross Timbers. The Wichita Mountains boast a biotic remnant of mixed-grass prairie, thickening into Cross Timbers oak and western plains brush. Their fire-preserving pasts differ as well, a matter of scale. The Flint Hills segued into a style of private ranching that favored spring burns before bringing cattle in, and public holdings have been acquired in patches over the past few decades, a chronicle that places it with the eastern United States. The Wichita Mountains refuge is more than seven times as large as the Konza Prairie and has never been in private hands, following instead a western lands scenario through a series of federal designations.

What they share is what matters, however. In their unique ways they both resisted the plow, kept fire, and became regional landmarks for fire management. Yet in true east-is-east and west-is-west fashion, they diverge on the character and frequency of burning. The Flint Hills, dominated by private landholdings, burn on regular schedules, often annually.

The Wichita Mountains, a public wildlife refuge, complete with a wilderness area, burn more episodically and, in 2011, on a landscape scale that swept over most of the refuge.

The Wichita Mountains are a miniature of the Southwest stuck on the plains as the Black Hills are of Northern Rockies. The mountains' history is synecdoche for southwestern themes: the lingering Indian wars, the lateness of settlement, the reservation of lands into the public domain, an economy of tourism and federal investment. Where they diverge is in their fire history. The Wichitas weren't opened to laissez-faire overgrazing, they kept their grasses, they held to fire.

In 1869 Fort Sill was established on the southern fringe of the Wichita Mountains to secure a frontier that, during the Civil War, had become a lawless borderlands. In 1874 the local tribes—southern Cheyenne, Kiowas, and Comanches—launched the Red River War. When Quanah Parker's Quohadi Comanches surrendered at Fort Sill in June 1875 the southern plains wars ended. The Southwest's continued. Interestingly, POW Chiricahua Apaches were later transferred to Fort Sill, along with Geronimo (who was buried there). Meanwhile the mountains remained part of a Comanche-Kiowa-Apache Reservation.

Plans to open the reservation to settlement prompted President McKinley to proclaim most of the Wichita Mountains as a forest reserve in 1901, the same year the town of Lawton was founded on the newly available lands. Four years later President Roosevelt rechartered the site as the Wichita National Forest and Game Preserve. Bison enthusiasts immediately seized on the possibilities for restoring the nearly vanished species, and in 1907 the New York Zoological Society delivered six bulls and nine cows, which also attracted tourists. Oklahoma's first tourist resort, Medicine Park, was platted the next year on the eastern slopes of the Wichitas. An elk reintroduction program commenced. Mountain reservoirs soon followed. In 1927 Congress authorized the purchase and relocation to the preserve of 30 Texas longhorns, by then a feral relic species more than working livestock. The CCC remade the mountains with roads, small dams for wildlife, and a game-retaining boundary fence. In 1935, recognizing that wildlife, not forests, was the thrust of management, the national forest was transferred to the Biological Survey and absorbed into the archipelago of wildlife refuges. Tourism added another feature with the Holy City pageant.

All in all a western history and a western-state geography. The refuge spans most of the mountains, flanked by a military and tribal reservations to the south and private lands to the north. The regional economy is mixed, but its infrastructure has been built with federal funds, and it relies on military, tribal, and tourist dollars. The Wichita Mountains refuge attracts 1.5 million visitors a year. About half the refuge's lands are in a special-use area with restricted access that includes a research natural area and two legal wildernesses. On its western flank it holds the Charons Garden Wilderness; on its east, Lake Elmer Thomas Recreational Area and Lake Lawtonka, both reservoirs that serve Lawton, and of course Medicine Park. Its early protection and successful reintroductions of charismatic wildlife have given it an unusually complete complement of prairie fauna, probably rivaled only by Wind Cave National Park in the Black Hills. The refuge proper spans 59,020 acres.

Over the long decades the mountains and their administering agencies had redefined themselves and their mission. The Comanche Nation has its tribal headquarters on Bingo Road in Lawton. Fort Sill converted from cavalry to field artillery. The Wichita forest reserve evolved into a premier wildlife refuge. And so too landscape fire has changed but endured.

Through rhythms of drought and soaking rains, across a history in which peoples have arrived and departed, often violently, over two centuries in which anthropogenic fire varied in frequency, timing, intensity, and biotic severity, flames persisted on the mountains. Its fire-scarred oaks record a mean fire frequency of 3.4 years prior to American settlement and 4.4 years afterwards. Lightning accounts for a scant 2 percent of ignitions.[2]

Dating back to 1712 yields a long fire chronicle but not one that transcends the introduction of the horse, the revolution in relations among the plains peoples, or the teleconnections with European settlement (for example, by disease). A baseline must appeal to basic biology, the rule-of-thumb reckoning that mixed-grass prairie burned every three to five years. Places with less frequent fires or not pocked with high-severity episodes will soon find themselves overrun with trees. Within that chronicle waves of fire rise and fall. A major outbreak occurred when the worst drought in the 300-year record coincided with the American Civil War and the decline of bison such that the prevailing ecological and political controls broke down, and fires, predictably, swept up the pieces. The indigenes even employed a scorched-earth strategy to discourage outside herders and horsemen. Gradually, as domestic cattle replaced the bison, burning continued from other, more benign causes. When the reserve was formally gazetted, the record marks a weakening in fires and a strengthening of woody species. In all this there is little outside the received narrative of fire in the American West.

The national forest was grazed, as was most in the West, and fires were fought, as again was the norm. But the Wichita never had the resources to fight big or fast fires, lacked the imperative to protect timber and watersheds, and was not obligated to provide winter forage as big ranches were. Instead they adapted local mores and what was termed a "defensive" posture toward fire. Staff relied on grazing to hold down wildfires; and when fires did break out they were inclined to pull back to a defensible road or granite ridge and burn out. Grazing did not penetrate deep into the mountains, so there was always some fuel to burn; and when the forest became a refuge and cattle were removed (save for the relic longhorns), still more grasses were released. By keeping herds within prescribed limits the refuge recreated a simulacrum of presettlement conditions in which grazers and flame competed. There was enough for both, but not too much. When the great drought of the 1930s hit, fires did not break out as they had in the 1860s. Not only had the refuge rebalanced its fuel matrix but it also was no longer a war zone. It had kept fire on the land, not by igniting it, but by tolerating it when it occurred and granting the resulting flames some room to roam.

What salvaged the Wichita Mountains, in brief, was what didn't happen. The torch had passed. During the 20th century, up to the fire revolution, the refuge's firescape contracted, but it did not shrink to infinitesimal slivers. Fire management had emulated wildlife management: once reintroduced (or allowed to remain), it had been left to free-range with some regular culling. As prescribed fire came into vogue, and as the Fish and Wildlife Service upgraded its fire capabilities to national standards, the refuge's fire officers began exploring how to boost fire's presence by calculated burning. This, too, tracked the trends in western public lands, with one critical difference. Fire in most places had been so downbeaten that it returned either fitfully or ferally. The Wichita Mountains had kept enough that they could notch up their regimen to something like historic proportions.

Still, the lessened pressure from fire had opened cracks for the extant woods to wedge open. Their savanna woodlands were thickening into forests, eastern red cedar was invading, and the old surface burns were inadequate to drive them off. Paradoxically (or perversely) the artillery ranges at Fort Sill, burning inadvertently (but inevitably) season upon season, year after year, grew some of best mixed-grass prairie anywhere. But of course the fort didn't have to fold in grazing fauna or maneuver around tourist browsers. Still, the contrast across the demilitarized zone demonstrated that the land could take a lot more fire and that the grasses would welcome a richer and more motley burning regime.

Over the past 20 years the fire organization moved from locals who grew up with the refuge to a full-staff cadre trained to NWCG standards and capable of interchangeability among other agencies. The program eased into a more formal mode that sought ways to bolster its burning and apply fire to beat back the encroaching woods. Experiments with thinning and burning, however, upset important constituencies (including administrators). Whatever its limitations the refuge's size, long history, and preserved fire argued for it to become a service center for the region. Its engines, crews, and knowledge support prescribed fire throughout refuges in Oklahoma and north Texas. Its research natural area has made it a darling of plains fire research.

Like its bison and elk, fire has returned, even if, like its resident longhorns, the final outcome includes a few exotics.

Whether or not the 2011 season should have come as a surprise, it did. An ice storm in January 2010 shattered overgrown woods and littered the landscape with downed debris. What ice smashed, drought blanched, and fire then seared with unprecedented severity.

Three fires affected the refuge, spanning a full season and the full extent of the preserve. The first, the Indiahoma Wye fire, began on the Fort Sill range and broke into the western edge of the refuge on April 12, where it scampered over 2,600 acres. The second, kindled on June 23, also bolted out of Fort Sill and, while it missed the refuge proper, ran into Medicine Park on its eastern perimeter and forced the refuge to respond. Before it quieted, the fire burned 3,000 acres and 14 houses. The monster burn, the Ferguson, waited until September 1, when repairs at a culvert on the road separating Fort Sill from the refuge threw a spark that instantly dashed beyond control. The flames burned north where they burst out of the refuge boundary before burning west and blowing across the refuge and out its western border, sparing only those portions that had recently been prescribe burned. At 39,907 acres, it was a landscape fire that seared nearly half the refuge and spilled out into private and Bureau of Indian Affairs lands. It burned through the Charons Garden and North Mountain Wildernesses, through the research natural area, past stock reservoirs, over picnic areas, and around the Holy City, over range ripe with bison, elk, and prairie dogs. The Ferguson fire merged with that constellation of big burns that ravaged the Southwest from the Huachuca Mountains of Arizona to the piney woods of east Texas.

Where grasses serve as fuel, large-area fires can return yearly. The Ferguson in particular simply moved too fast to contain, and so it sprawled farther than previous fires of record. Its distinguishing feature, however, was its sheer power. But surely fires of similar severity had occurred in the past, or the landscape would have been smothered over the centuries by creeping woods. What the 2011 season demonstrated was not merely the need to ramp up prescribed fire, but the obligation for some high-intensity burns among that mix, fires that were severe but not savage.

The refuge's fire scene is a western one but with an Oklahoma accent. In most of the West fire officers lost the capacity to intervene. The grasses went, and with them went the ability to easily start and stop fire, and to rearrange its regime. While, during its century as a public estate, the Wichita Mountains have lost ground—have had less fire than they need, have picked up too many trees, have dampened the high-intensity outlier burns that it appears did the harder biological work—they kept flame. In all this the fire character of the refuge shuffled westward, but with a difference.

Aldo Leopold once famously noted that the first rule of intelligent tinkering is to save all the pieces. The Wichita Mountains did this on a large enough scale to allow those biotic springs and gears to be reassembled into slightly different forms. By keeping fire it kept the future.

THE BLACKENED HILLS

THE BLACK HILLS are an island, a cameo of the Rocky Mountains plucked onto the plains. In their geology, their biota, their history, their contemporary economy, the hills align with the West. But its fire regimes seem to have absorbed some features from the surrounding prairie. It's as though, here, the plains congealed into a patch of the Rockies.[1]

The mountains bring rain. Like an ocean isle, the moisture is greatest on the windward (northwest) side and lightens to the lee (southeast). The rain grows forest amid what would otherwise be mixed or shortgrass prairie. But while the trees come from the Rockies they behave like brush on the plains. The western yellow pine more resembles eastern red cedar than it does the ponderosa pine of montane Montana or the Mogollon Rim. It propagates like a woody dandelion. It regenerates spontaneously, gregariously, promiscuously: only the saw, beetles, and flame hold it in check. Save that a timber market exists for the larger trees, and that its massed forest serves as an amenity for the tourist industry, the pines might well be considered a noxious woody weed. And like the cedar their history is intimately intertwined with fire, which is to say, with how natural conditions and human history have met, mingled, clashed, and colluded.[2]

The anomalies begin with their geographic location, as an outlier of the North American cordillera. What Midway is to the Hawaiian islands, the Black Hills are to the Rockies.

Around the northeast border of Colorado, the central Rockies thin and pivot to the northwest. The Black Hills continue the old trajectory, like a Pacific island stranded by shifting plates, or in this case by the shifting rumbles of the Laramide revolution. Like such isles the hills are igneous: their nucleus is a bubble of granite that rose and bulged the surrounding strata, and then abraded down. The overlying and the peripheral weaker layers washed away, leaving a hard core and a crusty ring like a barrier reef. Both peak and reef were forested, while a lagoon of grass lapped between them. The resulting dome resembled an etched turtle shell. Even with thousands of feet of rock removed Harney Peak remains the highest point in the United States east of the Rockies. The Black Hills rise out of the Great Plains like Tahiti out of the Pacific. From every vantage point around it attracts the eye. It draws all to it.

It's a biotic island as well. Remoteness, however, can simplify as well as retain. Being more mobile the hills' fauna could reach the isle from elsewhere as well as remain while the post-Pleistocene climate staggered toward some degree of stability; its complement of mammals is astonishingly rich. The flora is simpler. The grasses of the plains wash against the slopes and wend through the interior valleys. The trees are residual, however, a forest ark from the Pleistocene. For the Black Hills National Forest, which largely excludes the grassy valleys, some 95 percent of its land is forested, but 93 percent of that is a single species, *Pinus ponderosa*. The other dominants typical of the Rockies are absent, save aspen and spruce; these claim niches, ecological relics like the mammoth bones in sinkholes outside Hot Springs. The informing principle of the biota is its exuberant pine, which binds everything else into a whole.

The hills both drew and held that most peregrinating of creatures, humanity. The colonizers collected and concentrated almost every theme of western settlement. Their story, if staggered slightly behind most regions, claims it all in miniature: a sprawl of mining, railroads, mass grazing, widespread logging, and, after the wreckage became unbearable, late reservation to public lands and administration under doctrines of state conservation. The Black Hills Forest Reserve was proclaimed by President Grover Cleveland in 1897, the year the forests received their organic act and 23 years after Custer's Black Hills Expedition set off a gold rush. The reserve's boundaries encompassed most of the hills that lie within South Dakota, but those borders incorporated swathes and packets of private holdings from mining claims and ranchers patenting water holes, scattered across the mountain like beetle borings. The U.S. Forest Service assumed control in 1905. Seven years earlier, Henry Graves, a future chief, had visited the hills and despaired for their future. Their woods, their waters, their very soil was disappearing as though down a sluice box.

Gradually, the USFS wrested the region away from its ruinous laissez-faire rushes and imposed a minimal order. It brought some regulation to logging, forest grazing, watersheds and reservoirs, and of course it imposed fire protection. Still the Black Hills were, and have remained, overwhelmingly the dominant economic engine of the West River country, the impoverished stepchild and federal-laden half of a lightly settled state. (South Dakota is the 17th-largest state but has one fourth of one percent of the nation's population.) The Old West economy of commodities spiky with booms and busts hung on surprisingly long. The Black Hills National Forest continued as catalyst and counterweight.

Eventually the economy of the hills, as with the West overall, turned to services and amenities—to summer homes, recreation, and tourism, much of it playing on an idealization of that abusive past, as though the Hickoks and Oakleys of logging and grazing had become tradable securities in bidding for public attention. Eventually the era of hard-rock mining symbolized by the Homestake yielded to the chiseled granite of Mount Rushmore. Still, whether as emblem or economic player, the old order persisted. The ponderosa pine continued to hold the parts together. The Black Hills remained a living keepsake of the Old West, preserved, as it were, in pine pitch.

In eerie ways, the Black Hills came to resemble Lake Tahoe. One was a hump on the plains, and the other, a hole in the mountains, but they shared a comparable power to distill and concentrate the themes of western history. Each created a bounded miniature. Each is a lodestone for its regional economy, and its evolution. Both began as commodity producers before segueing into a modern economy of recreation and service. Without a governing body to orchestrate use on all lands, the hills, however, have boosted schlock to an order that could make Tahoe blush. Its private lands are littered with the bric-a-brac of industrial tourism like beer cans tossed along a highway, a cacophony of niche theme parks, kitsch, specialty museums, and souvenir shops like baited hooks for passing cars. But where Tahoe centers on a lake, and overseers are obsessed with its purity, the hills have an inland sea of woods, fecund to excess. Cutting in the Tahoe Basin practically requires an act of Congress. Cutting in the hills is essentially an act of survival.

The Black Hills have retained an old-model multiple-use forest of the sort that has become extinct elsewhere. It is both a working and a heritage forest. It stands to the national forest system as the rebuilt Deadwood does to mining; its evolution is like watching a frontier dance hall morph into a modern casino. Whether its continued high-volume logging is a salvation or an administrative equivalent to a bank of slot machines, generating extra revenue but at the cost of what Hal Rothman, historian of western tourism, termed "devil's bargains" remains to be seen; the most likely outcome may be both. Revealingly, the region has a state-legislated Black Hills Fire Protection District that aims to regulate all open fires and to attack any that occur.

What seems clear is that the Black Hills concentrate what comes to them. The good news is that they thus overcome the plains' tendency to disperse. The bad news is, they concentrate the bad as well as the good.

Their fires are a hybrid: more powerful and persistent than prairie fires, less consuming than the classic Big Burns of the Northern Rockies. Fires can spread at rates more typical of grasslands: the Jasper fire clocked off 30,000 acres in six hours. But thanks to pockets of heavy duff and abundant dead-and-down trees they can hold for five to seven days, well beyond the one or two burning periods common for even monster grassfires. The pine, that is, behaves more like prairie brush than either woods or grassland alone.

Intriguing facts—but the big story is the magnitude of burning that must have shaped the presettlement forest. Thanks to W. H. Illingworth, who accompanied Custer's 1874 expedition as photographer, a documentary record exists of what the Black Hills looked like a century after the Sioux laid claim and on the cusp of an American invasion. Grassy valleys are free of woods; south-facing slopes are culled of dense forest, letting trees flourish in stringers and pockets; and north-facing slopes are dappled with aspen and pocked with canopy holes. A century later those same scenes were rephotographed with astonishing results. Despite being logged over at least once, despite waves of bark beetles descending on the pine like locust swarms on corn, despite serious fire years, the forest had thickened at least threefold. South-facing slopes filled in. North-facing slopes shed their mottled texture in favor of a uniform canopy of even-aged pine. The only explanation for such a dramatic shift is an overall reduction in fire. The amount of presettlement burning—surface, crown, mixed; spring, summer, fall—must have been staggering to fight back the woody brush.[3]

The hills, while making a compact unity when compared to the surrounding plains, are a complex of fire regimes when examined in detail. Chronologies of fire-scarred trees suggest long waves in which burning overall waxed and waned with climate and human migration. Average return intervals are 16 years, with some sites as frequent as 2 years and others reaching 34 or so. The serrated texture of the geologic dome that defines the hills offers plenty of room for variability; averages across the hills probably mean little. Moreover the fuel arrays defy simple cycling because they are mediated by two fauna. Grazers, most spectacularly the bison, compete with fire for the grasses, while bark beetles work over the pines. What the record clearly reveals, in common with western landscapes generally, is that the old regimes broke down in the late 19th century. Somewhere between 1881 and 1893, the last landscape fires occurred. Over the next century big fires meant a single burn in the occasional year on the order of 10,000 to 20,000 acres.[4]

The bill of indictment cites the usual culprits. Logging shattered the structure of the old forest, grazing checked the grasses that might have helped keep fire on the ground, and the Forest Service committed to fire's exclusion. A classic photo of light-burning was taken in 1898 showing Henry Graves, the man who would lead the fight as chief forester, standing among Black Hills pine and smoldering surface burns. The exuberance of pine regeneration made a mixed fuel array not only the norm but a constantly primed tinder box. Having proclaimed the forest a poster child for all the ills of laissez-faire land use, the Forest Service had to demonstrate an alternative. It could not abolish mining, logging, or grazing; or throw out the thousands of homestead patents; or wave off mountain pine beetles. But it could attack fire.

It had modest success, perhaps from fire's exhaustion as much as its suppression. Then the Great Drought framed the hills with big fire years in 1931, when fire burned to the Rochford city limits, and 1939, when the McVey fire roared outside Hill City; three years of flooding followed. Between those dates the Forest Service adopted the 10 a.m. policy. The state legislature responded in 1941 by creating the Black Hills Fire Protection District to regulate all open fires. In 1959 a fire burned to the edge of Deadwood. In 1960 two large fires broke out near Wind Cave, and in 1964 the park experienced its biggest burn, the Headquarters fire, at 14,096 acres.[5]

The modern era commenced in 1985 as a new spiral of drought began circling the West. Big fires broke out in 1988 and again in 1990. Then the Jasper fire—part of the 2000 Northern Rockies constellation—ripped over 83,511 acres and burned a cavern in the center of the Black Hills. Thereafter large fires have struck almost annually. The southern ridge—the reef around the island peak—have caught the most; Elk Mountain suffered so many repeat burns it was no longer even black. Fire prevention succeeded in holding the number of human ignitions; but lightning, like coyotes filling the void left by vanished wolves, more than made up the difference. Then drought and beetles began mutating green forest into red slash and standing fuel, and brushed against a built landscape that, morphing into amenities communities, magnified the risks to life and houses. The forest of the Black Elk Wilderness upwind of Mount Rushmore is 85 percent bug-killed. Instead of sprawl, the Black Hills suffered from infilling—helpful in cities but dangerous in wildlands. The collective landscape threatened to sink into a forested variant of gambler's ruin.

As fire risk went ballistic, fire protection scrambled to keep pace. Spooked by the Jasper fire, the various fire authorities in and around the Black Hills established the Great Plains Interagency Dispatch Center at Rapid City. In 2002 an executive order by Governor William Janklow consolidated the authority of the state forester over the Black Hills Fire Protection District. In 2006 North and South Dakota, Wyoming, and Colorado signed the Great Plains Interstate Fire Compact. The thrust to all these efforts was protection.[6]

The Black Hills National Forest ran a full-gamut suppression program. From 1982 to 2011 the forest annually averaged 120 fires that burned 9,709 acres; over the past decade, aggressive fire control brought those numbers down to 97 fires and 6,356 acres a year.

Behind both landscapes, the built and the wild, lie the ponderosa pine. It underwrote the paradox that while people sought to save the forest, they also had to save themselves from it, for even as fires and beetles took out mature stands, seedlings sprouted like ragweed. The forest seemed to grow faster than it could burn.

It was too dangerous to let wildfire and free-ranging beetle infestations remove the trees, not least because both left abundant fuels in their wake. The dead trees, once fallen, could stoke pilot flames into the maturing thickets of pine regrowth. Instead, in an almost textbook example, the national forest turned to silviculture. It would let prescribed felling control fire, and then let prescribed fire assist continued felling. Whether cut, chewed, or burned, the pine seemed irrepressibly capable of regenerating, and only some kind of thinning program could prevent the revanchist woods from consuming the landscape. To the mind of the public forester the agency needed to regulate nature's laissez-faire economy as it had that of the settlement era.

The forest has turned to cutting to fight off beetles—sending mature timber to mill, cutting and chunking where patches are too tiny for commercial harvest, thinning ahead of infestation frontiers. It cuts for similar purpose around towns, monuments, and I-zones in the name of fire protection. It logs to create a mosaicked landscape, one better buffered against wildfire. It relies on variations of shelterwood thinning, not clear-cutting, so seed trees remain. In the 1990s the forest turned to whole-tree logging, which further diminished the visual impact of large harvests. The land appears to heal rapidly. Regrowth soon covers what traces remain. There is virtually no public protest.

Instead of holes, however, there are piles. A market exists for mature trees; none for the smaller diameter stems and tops, which are stacked into mounds the size of strip malls. Here is where prescribed fire enters the cycle: when snows prevent spread, the piles are ignited. Some broadcast burning occurs, but mostly burning serves cutting, just as cutting is intended to assist the control of wildfire. But even when stacked like cordwood (literally) the burning falls behind. After a dry winter or two in which firing is omitted the Black Hills might better be called the stacked hills. In the meantime an immense network of roads laces the mountain that allows for rapid initial attack.

In principle it is a forester's dream. But logging hasn't stopped the beetle outbreaks every 30 years or so; the hills have been logged over several times, and the beetles keep coming. Nor has it stopped the fires; the roads, the thinning, the whole-tree harvests, none have prevented ignition or scotched the prospects for further Jasper-sized burns. The Jasper fire, after all, burned through landscapes logged several times. The past decade has even experienced crown fires at night, unheard of before the modern system came into sync.

At the most basic level the program doesn't go far enough. Timber managers estimate they need 50,000 acres treated a year, and fire managers want at least 35,000 acres thinned and burned. Timber operations achieve about 5,000 acres, while fire has achieved about 4,400 acres annually over the past 10 years. A decade ago the Black Hills surpassed the collective national forests of Region 6 in timber production; now it exceeds Montana, and it achieves that output not with big trees but through the sheer volume of its smaller ones. The market for piled fiber, however, is very small. Without extraordinary effort, the prospects are that regeneration will romp through the hills, and that fire will rip through the regrowth.

With 3,500 miles of internal borders due to private inholdings, and with prospects for explosive runs, there is little opportunity to allow landscape fires much room to roam; the developed sites all seem to lie downwind from the relatively unbroken forests. For the same reason, it is tricky—more art than science—to finesse and coerce prescribed fire to behave as desired and stay within its designated bounds, like training a grizzly bear to dance or a mountain lion to cuddle. For a forest that struggles to find the prescription windows to keep up with its pile burning and the operational latitude to conduct broadcast burning on the scale required, fire management must veer toward suppression and the hope that, between beetles, saws, and burns, fire officers can gain enough space and sufficient time to catch the bad fires before they blast into communities.

As it approached the 2012 season the forest could certainly boast an enviable fire establishment. It had 14 engines, 2 hotshot crews, a 7-person helitack crew, a type 1 helicopter, a tanker base at Rapid City, access to 3 dozers, and a peak-season staff of 150. The Black Hills Fire Protection District and Great Plains Interagency Fire Coordination Center gave it an institutional context for cooperation. The forest was the largest player by far in South Dakota. It had been fighting fires for over a century.

In triangulating the bulk pyrogeography of the Great Plains, the Black Hills suggest two contrasts. One is to the eastern prairie, one to the western forest. The hills are to the western plains what the prairie peninsula is to the eastern, putting trees amid grass instead of grass amid trees. The one puts the West into the plains, and the other, the plains into the Ohio Valley. What this means beyond an interesting intellectual exercise is unclear. What makes the Black Hills significant to wildland fire is that they display an alternative history, what multiple-use fire suppression might look like had the nation chosen to stay on the old slash-and-suppress path during the 1960s.

Like a vintage car, the old model keeps being rebuilt and upgraded as it adds mileage. The forest has advantages not generally available. There are no threatened and endangered species to allow environmental groups leverage; only one token wilderness exists, the Black Elk, with an area probably equal to that of the Rushmore and Crazy Horse monuments; and the local communities support the timber industry. Two-thirds of the forest is available for harvest and accessible by road, and salvage logging is common. In a state of 824,000 people, almost all of them east of the Missouri River, the Black Hills remain a gold mine, even if it means panning for tourists rather than nuggets. Most of the classic tools of forestry still line the workshop walls.

But if the Black Hills National Forest concentrate the opportunities for continued multiple-use management, it also concentrates its ills. Summer homes sprout like mushrooms but do not make a sustainable economy. Logging cuts more or less freely but cannot alone halt beetle epidemics or megafires, and leaves mounds of debris like woody junkyards and an economy that continues to decline. Deadwood was reborn as a casino city and TV celebrity but continues to wither nevertheless. Grazing persists unimpeded, jostling against fire and wildlife, yet it remains economically marginal. Drought drains streams along with the Pactola Reservoir. In 1984 the BNSF railroad shut down its last line through the hills. The mines that established the cycle of boom and bust, after a brief bubble, have collapsed once and for all. Galena lives on as a superfund site. The Homestake Mine finally shut down and bet its future on hopes that the National Science Foundation might turn its monumental shaft into a field lab for neutrino research. One can imagine no better illustration of the shift in the hills' fortunes.[7]

And fire? The Black Hills may evolve into another kind of experiment: a test on the ability of the multiple-use forest to achieve a level of fire management that has eluded (or been denied to) other models of national forest administration. The complexity of fire management here simplifies. New starts are suppressed, prescribed fire is mostly pile burning, and the landscape links flame to fuel, not to matters of ecological integrity. Legal, social, and political shackles that fire managers elsewhere blame for making their job impossible are here unfettered. Instead the complications reside in economics: in dispersed summer homes and tourism, weak markets, and the costs of remoteness. The isolation that makes the hills special also renders them vulnerable. The hills concentrate fuel and fire problems without allowing an equally focused response. The fires come, and they threaten to rush over the land with unremitting frequency and with perhaps greater severity.

Old-style forestry claimed it could manage fire if silviculture was granted carte blanche and fire suppression staffed adequately. In the Black Hills it has been granted that wish probably as fully as any national forest in the West. What happens will thus affect not only the hills themselves but the ceaseless national conversation about why wildfires continue to metastasize and what might be done, at what costs, to stop them, and what of those fires might need to endure, even if a final reckoning may conclude that fire here is no worse than in other places, and no better, that it is simply another anomaly rising out of the plains.

EPILOGUE

The Great Plains Between Two Fires

We must face the fact that the splendid story of the pioneers is finished, and that no new story worthy to take its place has yet begun.[1]

—WILLA CATHER (1923)

THROUGHOUT THE GREAT PLAINS the story of settlement carried the story of fire in its Conestogas and saddlebags. Prairie fire was a wonder and a terror: it was nature sublime. Over and again it appeared as an indelible feature of the American steppe and an inevitable chapter in the retold sagas of colonizing the land. In the end it reenacted the tragedy of the pioneer, in which the act of pioneering destroyed the conditions that had made pioneering possible. The fires mostly went away.

As the land filled out with farms and ranches, as isolated homesteads thickened into villages and cattle drives were penned into feedlots, the fires passed away along with bison herds and mounted tribes. In the southern plains it was fought and eventually crushed by intensive grazing and farming. In the northern plains it was squeezed out of the land by enduring and cultivating it away and replacing it with tame species. It survived in places that because of stone, sand, or government edict had resisted the plow and kept the old ways. Wildfire receded along with grasshopper infestations. Controlled burning became a memorial constructed out of natural materials akin to those plaques, museums, and replicas with which subsequent generations filled the countryside to honor the pioneering age. It became a relic practice or a lost fragment of prairie that had somehow been spared the full brunt of settlement.

Today, for reasons both unavoidable and promoted, flame is returning, not unlike mountain lions slowly expanding their range over the Sandhills and even into Sioux Falls. Where fire had survived, notably the Flint Hills, deliberate burning is modifying its regime, reinstating old rhythms much as it is reintroducing former fauna. In the southern plains it is reappearing mostly in feral forms. In the northern plains its return is mixed, as fire kneads into the dough of working landscapes. To date, the quota of wild fire is proportionately large, and of tamed fire, tiny. Outside the southern plains fires appear with the anomalous rhythms of measles outbreaks in places that failed their vaccinations. There is little, however, to interrupt the trend toward both wild and prescribed, both of which are likely to grow.

What is missing, however, is what Willa Cather foresaw almost a century ago. The new fire scene does not fit into the old fire narrative tied to settlement, nor does it connect to the national narrative. Like its migratory herds and flocks the story moves north and south, not east and west. Fire remains a peculiarly local practice, like county historical societies, or it exists as a subset of the national story, downloaded through federal agencies. The restoration of fire has followed, not led, national reforms. The plains seem destined to be a place where larger themes pass through and only occasionally or locally pause and send down roots. Probably only Texas will invent a distinctive narrative. For the rest, fire on the plains will be a place where the national story flies over or rolls across.

How those national trends actually play out of the Great Plains will likely depend on the particular interplay between the region's tendencies toward the fixed and the fluid. In the 19th century the means of stabilizing depended on barbed wire and windmills, as a way to halt migrations and to bind surface needs to subsurface sources. In the 20th, they pointed to transmission wires and gas wells, as a way to hold scattered communities together and to fasten a shifting surface to subsurface aquifers of oil and water. Those deep holes act like tent bolts to anchor the countryside against the trends to disperse.

In 1869 Currier and Ives published a famous lithograph to commemorate the completion of the transcontinental railroad. *Across the Continent* neatly divided the American continent between wild and civilized. The wild had no history; the civilized tracked a sequence of pioneering from explorer to schoolhouse. The tracks race across the plains to the western mountains.

Three years later Currier and Ives published another color print of a locomotive on the plains. *Prairie Fires of the Great West* shows a train, its smokestack belching ash and embers, racing from right to left, or if one superimposes a map (as most readers would do) dashing from east to west; it is, literally, an engine of settlement. This time the tracks divide an unburned foreground from a background of blackened prairie, flames wildly gusting, driving bison before it. The flames and fleeing fauna are dashing in a direction opposite the locomotive. The iconography is simple enough. The train is moving west into the future. The prairie fire is moving into the past.

Considered as two forms of combustion, open burning of prairie and internal combustion of coal, the lithograph depicts perfectly the transition that has defined the settlement era as it has sought ways to fix itself to the landscape. Fossil fuel supplied power, substituting for the lack of wood; it powered the manufacture of barbed wire and windmills, which helped stabilize life on the plains; and it stoked the locomotives and later 18-wheelers that bound the economy of the plains to the national market. On the Great Plains, as everywhere, the contemporary scene reflects the competition between these two grand realms of combustion. Free-burning fire passed from the scene; internal combustion drove into the future.[2]

To those images add this: a succession of BNSF trains pulling hundreds of coal cars through the Sandhills, burning diesel to haul fossil biomass from Wyoming to Nebraska power plants, running parallel to a highway of automobiles powered by gasoline. If we wished a contemporary Currier and Ives scene of fire on the prairies this would be it. Those trains and cars are carrying a latent fire over the countryside. They are the equivalent to the long, blackened streaks of burned grass described in *Old Jules* and countless settler journals.

As settlement struggled in the western plains to counter aridity and drought, and to convert from ranching to farming, it sank wells to tap into groundwater. Underlying the central plains is, in fact, the largest subsurface pool of water in the country, the Ogallala aquifer. Some parts of the Ogallala are replenished by surface sources, but most is a Pleistocene relic, a vat of fossil waters. Tapping those waters has allowed (for a while) a stable settlement of the surface landscape; it countered the need to move or the disheartening tendency to settle, abandon, and reclaim. But the aquifer is emptying. In the end it will prolong the rhythms without changing their dynamic.

Other buried reservoirs have replaced it. Among the largest deposits of fossil biomass are the oil and gas reservoirs in the southern plains, to which in recent years have been added immense stocks of coal in the northern plains, and oil and gas in North Dakota. Those fossil fuels stake down the two termini of the Great Plains. They sustain the regional economy; they anchor north and south and allow transit east and west. Eventually these reserves too will exhaust, and the region will have to devise some other means to fix what wants to flow. For now those wells and mines have redefined the competition that locomotives and wildfire exhibited on the 19th-century prairie. But they have done so without replacing the iconography and narrative by which one can parse those Currier and Ives lithographs of life on the plains.

Those places that in fact kept fire did so because they stood outside the inherited narrative. How to understand revanchist wildfire and prescribed fire within a combustion environment shaped by fossil fuel, in which oil derricks and transmission lines replace windmills and barbed wire, is not yet written into burn plans. The wild fires of the new plains may be the gas flares that make western North Dakota among the hottest pixels in North America.

Reinstating fire—the right kind of fire—is not easy. Today's torchbearers resemble the isolated homesteaders and ranchers who sought to stay put in a hostile country. They will survive by enduring, adapting, and gathering into communities. They will bequeath diaries and stories and photographs that their occupational grandchildren may well honor with plaques and memorials if they can be sited within a compelling narrative and show how the region binds to larger, national themes.

A final paradox: the Great Plains are among the most mobile of American landscapes and among the most volatile. Perhaps only New England has undergone so radical a churn of countryside on this scale. One can find a relic sod house next to an abandoned Minuteman missile silo adjacent to protected Badlands and a national grasslands acquired from failed ranches. Schemes can come quickly, and just as rapidly fall apart. The state that originated Arbor Day is now trying to burn off invasive juniper. Almost as soon as the national government disposed of the land it began reacquiring it as the landscape resisted, homesteads went bankrupt, and the countryside began a slow, painful depopulation. The apparent tabula rasa of the terrain and its capacity to fill, overflow, and empty over and again argue that it can be easily changed for good or ill, and changed with the pace of prairie flame. There are opportunities for reform that do not exist elsewhere on a public estate that is hemmed in by political rivalries or on intensively cultivated private lands.

The most spectacular proposals involve returning the western plains to something like their presettlement condition, beginning with a restocking of native wildlife. In 1987 the idea was floated to rationalize and to place under federal guidance the reorganization of those lands as their relentless depopulation freed up land. The Buffalo Commons, as its advocates termed it, would involve 139,000 square miles across 10 states. The upshot would be "the world's largest historic preservation project, the ultimate national park." In effect what the nation (and federal government) had done in an earlier age by promoting small private landholdings, it would undo.[3]

The scheme went nowhere and managed to enrage regional powers and many local communities. But in truth the process was well underway. The federal government had already acquired grasslands, wildlife refuges, and national parks, and held in nominal trust considerable lands as reservations for Native Americans. Almost all such lands were in the northern plains. But government involvement included agricultural and other subsidies as well; the threat to property rights that aroused many landowners was already compromised by supplemental help that had been present since the Homestead and the Timber and Stone Acts. The deeper fury likely stemmed from the sense that the concept violated—in fact, inverted—the inherited narrative, which is what had given cultural meaning to those who lived on the land and justified what for many was an ordeal. As much as pivot irrigation and rural electrification, that story is what helped stabilize the plains' tendency to dissipate.

Outrage, however, did not halt the trends to depopulate and realign those western plains landscapes. A suite of proposals have since floated, many retaining property ownership or shifting the acquisition and management of land to private NGOs such as the American Prairie Reserve, which is attempting to rebuild native prairie on 3.6 million acres next to the Charles Russell National Wildlife Refuge in eastern Montana. Since 1992 an InterTribal Bison Cooperative has assisted Native Americans with restocking bison on reservations. The transfer of land to public ownership, or to private ownership toward a public good, is happening on small scales as well (for example with the Nature Conservancy). The future points to more cooperative projects.

Nor is the shift limited to the region's arid belt. Economic trends along with depopulation, or a population shift from country to city, are freeing up lands for repurposing. The Flint Hills are becoming as dense with preserves and conservation easements as any prime landscape in the country; only two involve federal agencies (the National Park Service and Fort Riley) directly. The Nature Conservancy continues to acquire premier sites. The Natural Resources Conservation Service and Conservation Reserve Program help set aside patches of farm and ranchland that are available for (and need) burning. Throughout the region there is more landscape and fuel available for burning and a pressing need to burn it deliberately rather than inadvertently.

Those newly available opportunities require some extra assistance if inhabitants are to replace feral fire with managed fire. Individual landowners cannot cope by themselves; the same argument that John Wesley Powell advanced in 1878, that the old patterns of settlement will not work in the "arid regions," holds for fire management as well. Inhabitants need capacity, which means collaboration, and often requires subsidies from some outside agency for land set-asides, burn trailers, or model bylaws. And they need an updated narrative, which must also come with some assistance, in this case intellectual boosting, from the outside. The inherited fire narrative, embedded in the story of pioneering, is rapidly being depopulated of practical meaning. It no longer explains how to cope with fire. If another narrative does not recolonize the land, the old one will continue to drift away, or find itself replaced by wandering packs of feral anecdotes.

To place the Great Plains in the national fire scene, look to their cardinal borders. Their north-south termini define their position within the grand story of fire on Earth. Their east-west borders inform their relationship to the story of fire in America.

What underwrites the first are the immense subterranean reserves of fossil biomass which anchor its combustion economy. The shift from free-burning fire to internal combustion has reconfigured landscapes throughout the country. In the plains those combustion aquifers have helped replace the subterranean waters that sustained an earlier generation's attempt to fix itself to the landscape and sweep fire away. An economy of industrial combustion, however, has worked to take both controlled flame and people out of the scene, which, paradoxically, is creating the conditions to bring free-burning fire back.

Its east-west borders describe the plains' connection to the nation. The issue has two parts. One, the Great Plains are an identifiable region in their own right with roughly 20 percent of the nation's landmass. They have characteristic biotas, distinctive fire regimes and practices, and a unique narrative to explain the historic relationship between the inhabitants and their land. A chronicle of fire on the plains differs from other regions as much as tallgrass prairie does from longleaf pine woods, or mesquite from cheatgrass. Fire management and history must cope with the region's oddities as much as fire ecology must explain its patch dynamics. The Great Plains are too big geographically and conceptually to ignore. The flow of fire history might dam up against the region, or flow around it, but cannot ignore it.

The second issue is just that: how do the plains integrate with the other regions or, to rephrase that concept, how do they help the rest of the country cohere or, to put it differently, how do they sit within the national narrative. Their contribution is peculiar in that, while geography suggests they are a keystone region, history does not. The Great Plains have not invented a master narrative of fire that has spread out from its core and bonded the other, peripheral regions to their and thus hold the nation's disparate pieces together; rather the opposite has occurred. It is a second act; its role lies in its mobility. It provides passage for the country's story to cross over, and in doing so, to undergo a phase change. To recall W. B. Yeats, the center cannot hold.

But then it doesn't have to. It only has to juggle its endemic tension between the fixed and the fluid, which historically has meant letting the big themes flow across, and from time to time to isolate and fix those features that speak to its own sense of itself. No pyric region has changed more from settlement, and none may change more in the future. America's national fire scene will change with it.

NOTE ON SOURCES

AS WITH THE OTHER VOLUMES in *To the Last Smoke*, my sources were mostly my study tours and conversations with hosts and guides. If the book were a fully fledged exercise in fire journalism, it would spend more pages profiling and quoting the fascinating characters I met. But federal funding meant my research fell under human-subjects guidelines, which was always awkward and often inhibiting. I had to rely principally on personal impressions and written documents. Again, as with the other volumes, I list the people and relevant primary documents under each particular essay. I offer in what follows some recognition of the written literature that I found most helpful across the region.

For historical conditions, see Julie Courtwright, *Prairie Fire*. For their own contributions to history, I would add Webb, *Great Plains*, and Omer C. Stewart, *Forgotten Fires: Native Americans and the Transient Wilderness* (Norman: University of Oklahoma Press, 2009), which brings together Stewart's early and oft-forgotten essays. On historic sources of ignition, see the still indispensable Higgins, "Interpretation and Compendium." The Missouri Tree-Ring Laboratory under Richard Guyette and Michael Stambaugh has been producing a constant stream of fire histories from sites in or around the plains—a unique chronicle.

Beginning with Wright and Bailey, *Fire Ecology*, the grasslands have sprouted a scientific literature both academic and popular. Among recommended volumes see, for an overview, Scott Collins and Linda

Wallace, eds., *Fire in North American Tallgrass Prairies* (Norman: University of Oklahoma Press, 1990); for an academic inquiry, Alan K. Knapp et al., eds., *Grassland Dynamics: Long-Term Ecological Research in Tallgrass Prairie* (New York: Oxford University Press, 1998); and for a popular introduction, Chris Helzer, *The Ecology and Management of Prairies in the Central United States* (Iowa City: University of Iowa Press, 2010).

NOTES

PROLOGUE

1. Washington Irving, *A Tour on the Prairies* (Paris: A. and W. Galignani, 1835), 5.
2. Edwin James et al., *Account of an Expedition from Pittsburgh to the Rocky Mountains* (1822; repr., Ann Arbor, MI: University Microfilms, 1966), 178. Saying borrowed from epitaph in Paula M. Nelson, *The Prairie Winnows Out Its Own: The West River Country of South Dakota in the Years of Depression and Dust* (Iowa City: University of Iowa Press, 1996).

THE FIXED AND THE FLUID

1. Phrase from Tim Flannery, *The Eternal Frontier: An Ecological History of North America and Its Peoples* (New York: Atlantic Monthly Press, 2001), 84.
2. Willa Cather, "Nebraska: The End of the First Cycle," in *O Pioneers!*, ed. Sharon O'Brien (New York: Norton, 2008), 335.
3. Edwin James et al., *Account of an Expedition from Pittsburgh to the Rocky Mountains*, vol. 3 (London: Longman, Hurst, Rees, Orme, and Brown, 1823), "Great Desert" on 276 and 285, quote from 24.
4. Walter Prescott Webb, *The Great Plains* (Boston: Ginn, 1931; Lincoln: University of Nebraska Press, 1981), 24.

GRASS

1. J. Evetts Haley, "Grass Fires of the Southern Plains," *West Texas Historical Association Year Book* 5 (1929): 24–25.

2. The literature on grasslands and fire is so large and so specific there is little point in identifying more than symbolic references. I found particularly helpful as a digest Chris Helzer, *The Ecology and Management of Prairies in the Central United States* (Iowa City: University of Iowa Press, for the Nature Conservancy, 2010). My figures come from this very practical and insightful summary.
3. William J. Bond and Jon E. Keeley, "Fire as a Global 'Herbivore': The Ecology and Evolution of Flammable Ecosystems," *Trends in Ecology and Evolution* 20, no. 7 (July 2005): 387–94.
4. Quote from Karen Smith from conversation with author, September 24, 2014.
5. Jefferson to Adams, May 27, 1813, quoted in "Thomas Jefferson on Forest Fires," *Fire Control Notes* 13 (April 1942), 31. Aldo Leopold, *A Sand County Almanac* (New York: Ballantine, 1970), 31. Carl O. Sauer, "The Agency of Man on Earth," in *Man's Role in Changing the Face of the Earth*, ed. William L. Thomas Jr., vol. 1 (Chicago: University of Chicago Press, 1956), 55. Sauer's essay on fire, 54–57, is a forgotten classic.
6. Haley, "Grass Fires," 43.
7. Julie Courtwright, *Prairie Fire: A Great Plains History* (Lawrence: University Press of Kansas, 2011), 127.
8. Courtwright, *Prairie Fire*, 130.

SEASONS OF BURNING

1. For a good survey of contemporary art, see Courtwright, *Prairie Fire*, who notes that the "beauty of the prairie fires thus became a significant piece of Plains identity" (162). The book has quickly established itself as a standard reference.
2. The saga of Bessey, Clements, Weaver, et al., is well summarized in Ronald C. Tobey, *Saving the Prairies: The Life Cycle of the Founding School of American Plant Ecology, 1895–1955* (Berkeley: University of California Press, 1981); for quotes, see 2–3. My survey seeks to place Tobey's admirable account within the context of Great Plains fire history.
3. Tobey, *Saving the Prairies*, 5, 2.
4. Tobey, *Saving the Prairies*, 3.
5. See Courtwright, *Prairie Fire*, 169–71, for an account of the South Dakota fires.

PLEISTOCENE MEETS PYROCENE

1. My debt is deep, but begin with the catalyst, Shane del Grosso. Then go to Ken Higgins, Pete Bauman, Jim Strain, and Kyle Kelsey, and on to Jeff Dion, Dan Severson, Cami Dixon, Arnie Kruse, and Karen Smith. They began my intellectual induction into a marvelous region.
2. For background references, I found useful Kenneth F. Higgins, "Interpretation and Compendium of Historical Fire Accounts in the Northern Great Plains," U.S. Fish and Wildlife Service *Resource Publication 161* (1986); Kenneth F. Higgins, Arnold D. Kruse, and James L. Piehl, "Effects of Fire in the Northern Great Plains," *Extension Circular 761* (Brookings: South Dakota State University, 1989), "Prescribed Burning Guidelines in the Northern Great Plains," *Extension Circular 760* (Brookings: South Dakota State University, 1989), and "Annotated Bibliography of Fire Literature Relative to Northern Grasslands in South-Central Canada and North-Central United States," *Extension Circular 762* (Brookings: South Dakota State University, 2000); Leo M. Kirsch and Arnold D. Kruse, "Prairie Fires and Wildlife," in *Proceedings, Tall Timbers Fire Ecology Conference* (Tallahassee, FL: Tall Timbers Research Station, 1973), 289–303; Todd Grant et al., "An Emerging Crisis Across Northern Prairie Refuges: Prevalence of Invasive Plants and a Plan for Adaptive Management," *Ecological Restoration* 27, no. 1 (March 2009), 58–65; Jerry D. Kobriger et al., "Prairie Chicken Populations of the Sheyenne Delta in North Dakota, 1961–1987," in *Prairie Chickens on the Sheyenne National Grasslands*, General Technical Report RM-159, U.S. Forest Service, 1987, 1–7; Arnold D. Kruse and Bonnie S. Bowen, "Effects of Grazing and Burning on Densities and Habitats of Breeding Ducks in North Dakota," *Journal of Wildlife Management* 60, no. 2 (1996), 233–46; Robert K. Murphy and Todd A. Grant, "Land Management History and Floristics in Mixed-Grass Prairie, North Dakota, USA," *Natural Areas Journal* 25 (2005), 351–58; and David E. Naugle, Kenneth F. Higgins, and Kristel K. Bakker, "A Synthesis of the Effects of Upland Management Practices on Waterfowl and Other Birds in the Northern Great Plains of the U.S. and Canada," *Wildlife Technical Report* 1 (Stevens Point, WI: College of Natural Resources, University of Wisconsin-Stevens Point, 2000).
3. Arrowwood refuge is fortunate in having a cache of narrative reports, and doubly so, in having digitized most. Not all years are included, but the detailed record gives a good sense of what developed.

4. Kirsch and Kruse, "Prairie Fires and Wildlife," 293.
5. For a summary of the controversy, so similar to those in America, see Stephen Pyne, *Vestal Fire* (Seattle: University of Washington Press, 1997), 361–64.

ECOTONE

1. Cody Wienk orchestrated an exceptionally productive visit, which Dan Swanson, Eric Allen, Amy Symstad, and Al Stover turned into a tutorial and field trip. The park is unusually rich in information, which the staff generously made available. As usual, how they see the land they manage is not how I see it, but interpretation is my charge—one I think an agency dedicated to public interpretation can understand if not appreciate. My thanks to all.

   The park has a wonderful library of data and reports. Among the most useful I found were the following: S. Bone and R. Klukas, "Prescribed Fire in Wind Cave National Park," unpublished report (October 27, 1988); Deane M. Shilts, "Experimental Burning in Wind Cave National Park," unpublished report (February 18, 1976); Peter M. Brown and Carolyn Hull Sieg, "Historical Variability in Fire at the Ponderosa Pine-Northern Great Plains Prairie Ecotone, Southeastern Black Hills, South Dakota," *Ecoscience* 6, no. 4 (1999): 539–47; and *Wind Cave Draft Fire Management Plan* (2005), which includes a good historical summary.
2. In 2012 the park acquired an additional 5,300 acres from a former elk ranch. These new acres are not part of the park's past but do belong to its future.
3. In April 2015 the Cold Brook prescribed fire escaped, burned a UTV, and became a wildfire that attracted unwanted attention from one of the state's senators. For a report on the escape and accident see "Cold Brook Escaped Prescribed Fire: Facilitated Learning Analysis," Wind Cave National Park, 2015, http://wildfiretoday.com/documents/ColdBrook_FLA_07_28_2015.pdf.

NIOBRARA

1. Thanks to Chris Rundstrom, Tracey Nelson, F. Richard Egelhoff, and Doug Kuhre for a gentle primer on fire and Niobrara. Thanks, too, for

the chance to wake up with the dawn's early light streaming over the Niobrara River.

2. Data from poster in Niobrara Preserve visitor center, from Thomas Bragg (June 2005). On the Sandhills to the south of the river, see Thomas Bragg, "Fire in the Nebraska Sandhills Prairie," in *Proceedings, 20th Tall Timbers Fire Ecology Conference, Fire in Ecosystem Management: Shifting the Paradigm from Suppression to Prescription*, ed. Teresa Pruden and Leonard Brennan (Tallahassee, FL: Tall Timbers Research Station, 1998), 179–94.

SANDHILLS

1. Thanks to Mary Lata for suggesting sites and contacts in the region and to Michael Croxen for a lively introduction to fire on the Bessey District today and for directing me to an unusually rich on-site historical collection. Most of the documentation and photos, not surprisingly, deal with the nursery. But fire was present, too, as its threat was ever present.

   Mari Sandoz, *Old Jules* (1935; repr., Lincoln: University of Nebraska Press, 1985), 3.
2. The most thorough account of these ideas is in Richard Grove, *Green Imperialism: Colonial Expansion, Tropical Island Edens, and the Origins of Environmentalism, 1600–1860* (Cambridge: Cambridge University Press, 1995).
3. On Bessey—Raymond J. Pool, "A Brief Sketch of the Life and Work of Charles Edwin Bessey," Papers in Systematics & Biological Diversity, University of Nebraska-Lincoln (1915); Richard A. Overfield, "Trees for the Great Plains: Charles E. Bessey and Forestry," *Journal of Forest History* 23, no. 1 (January 1979): 18–31; and Ronald Tobey, *Saving the Prairies: The Life Cycle of the Founding School of American Plant Ecology, 1895–1955* (Berkeley: University of California Press, 1981).
4. Charles E. Bessey, "Are the Trees Advancing or Retreating Upon the Nebraska Plains?," *Science*, New Series 10, no. 56 (November 24, 1899): 768–70.
5. Charles A. Scott, *The Early Days: The Dismal River and Niobrara Forest Reserves* (Washington, DC: U.S. Department of Agriculture, Forest Service, 2002), 34.
6. Sandoz, *Old Jules*, 300, 351–54.
7. Scott, *Early Days*, 29, 34–36, 42.

8. William G. Sullivan Jr., "The Plum Fire," Weather Bureau Airport Station, Denver, Colorado (May 27, 1966). Other fires from records on file at district headquarters.

THE NEBRASKA

1. I would like to thank Tristan Fluharty for organizing a useful meeting with the Nebraska National Forest and Grassland staff, who then directed me to some small caches of useful documents. Charles Butterfield of Chadron State University hosted a delightful field trip to sites along Pine Ridge. I would also want to thank Mary Lata for alerting me to the fire-management possibilities of the forest, for helping me make contact, and for a critical reading of a draft manuscript.
2. The best summary of relevant information are the forest's fire-management plans: the 2009, published, and the 2012, still in draft.
3. The story of the fire is well covered by two unpublished narratives, both in Nebraska NFG files: Marvin Liewer, "Ft. Robinson Fire Narrative," and Bob Sprentall, "Ft. Robinson Fire."

LOESS HILLS

1. An introduction to the Loess Hills was possible through the attention of John Ortmann, who prepared maps, documents, and a field tour (fortunately, a spell of rain had dampened the burning for my time on site). I would also like to thank John Brohman for allowing me to see his land and learn from his and his neighbors' very practical experiences.
2. Story from John Ortmann by e-mail.

KONZA

1. I want to thank Dave Hartnett for his help in setting up my brief visit, and Tony Joern for a wonderful morning's tour and tutorial; Tom and Barb van Slyke for smoothing my stay; and Tony Crawford and Kari Bingham-Gutierrez for rousing the Konza Collection from its lair in Kansas State University's special collections.
2. A rich literature surrounds the region. For a good introduction to the land, see O. J. Reichman, *Konza Prairie: A Tallgrass Natural History* (Lawrence: University Press of Kansas, 1987). For a sample of its folklore, see Jim Hoy, *Flint Hills Cowboys: Tales from the Tallgrass Prairie*

(Lawrence: University Press of Kansas, 2006). And since the hills form its emotional core, see Courtwright, *Prairie Fire*.

3. The best summary of Hulbert's life is "Proceedings Dedication, Lloyd C. Hulbert, 1918–1986," in *Prairie Pioneers: Ecology, History and Culture: Proceedings of the Eleventh North American Prairie Conference Held 7–11 August 1988, Lincoln, Nebraska*, ed. Thomas B. Bragg and James Stubbendieck (Lincoln: University of Nebraska Printing, 1989). Online copy available at http://images.library.wisc.edu/EcoNatRes/EFacs/NAPC/NAPC11/reference/econatres.napc11.i0006.pdf. Several curricula vitae that Hulbert created at different points of his career are available in Kansas State University Special Collections, Division of Biology, Konza Prairie, Accession Number U97.5, Box 12, File 1. These gave particulars about his CPS years.

   For an overview of the CPS smokejumper program, see Mark Matthews, *Smoke Jumping on the Western Fire Line: Conscientious Objectors during World War II* (Norman: University of Oklahoma Press, 2006).
4. Report of Committee, December 1953, Konza Prairie Collection, Special Collections, Kansas State University, Box 1, Folder 20.
5. The most thorough summary of Konza's evolution is Hulbert, "History and Use of Konza Prairie Research Natural Area," *The Prairie Scout* 5 (1985), 62–93.
6. On the speculation about the origins of Hulbert's interest in fire, see "Proceedings Dedication." For a digest of his scientific conclusions, see the posthumously published Hulbert, "Causes of Fire Effects in Tallgrass Prairie," *Ecology* 69, no. 1 (February 1988): 46–58.
7. Perhaps the best summary of its research richness is the application for a seventh term under NSF's Long-Term Ecological Research network, NSF 13–588. For this round the emphasis shifted toward climate: "Long Term Research on Grassland Dynamics—Assessing Mechanisms of Sensitivity and Resilience to Global Change" (Lawrence: Kansas State University, 2014).
8. From Lloyd C. Hulbert Memorial Service, Kansas State University Special Collections, Division of Biology, Konza Prairie, Accession Number U97.5, Box 12, File 1.

PATCH-BURNING

1. This essay is an exercise in reportage, the outcome of a too-brief visit in the spring of 2009 to Oklahoma State University at the invitation

of the Graduate Student Organization of the Department of Natural Resource Ecology and Management, which included a marvelous tour of the Nature Conservancy's Tallgrass Prairie Preserve and, unexpectedly, an outbreak of wildfire. Among the many persons who tutored me in the fire culture of the region, I would like to thank particularly Brady Allred, Sam Fuhlendorf, and Dave Engle of OSU, and TNC's preserve manager, Bob Hamilton. They were great hosts and even better teachers.

Since the topic deals with themes relevant to the Great Plains, I am including it with this volume. Its earlier composition accounts for its somewhat different style and emphasis.

TEXAS TAKES ON FIRE

1. Some of the ideas expressed in this extended essay reach back to my graduate school days at the University of Texas-Austin and some of the literature I read then in an attempt to understand Texas. Among them were Walter Prescott Webb's *The Great Plains* and Donald Meinig's *Imperial Texas*. For interpreting the historical origins of the Texas persuasion, many of their ideas remain relevant.

Among those who graciously allowed conversations or pointed me to pertinent literature, I would like to acknowledge the following in particular. I'm confident none would express their understanding in the terms I use, and should not be held liable for my recasting of their knowledge and insights.

On the Texas Forest Service: Tom Spencer, Rich Gray, and Bill Davis.

East Texas: Thanks to Randy Prewitt for a vigorous talk and tour about the Sam, along with a useful cache of supporting documents. I was able to supplement our discussion with a group conversation with Jaime Gamboa, Chet Dieringer, Bobi Stiles, and George Weick. I think it fair to say that while none of them would not frame the issues as I have, many of my ideas have come out of our visits, and I'm grateful.

North Texas: Special thanks to Jim Ansley and Mustafa Mirik for insights and a field tutorial on the inimical Texas mesquite.

West Texas: John Morlock and Bill Davis both kindly interrupted their routines to accommodate my queries and tolerate my assorted ignorance, and Larry Francell sent a thoughtful critique of my original draft.

Edwards Plateau: Charles "Butch" Taylor proved an indispensable font of knowledge, literature, and good humor.

Bastrop: Greg Creacy graciously accommodated my request for a talk. Jeff Sparks spoke generally about fire and the Texas Department of Parks and Wildlife.

USFS: For assistance on identifying the level of federal support, I wish to thank Lewis Southard and his staff, who surely had more pressing matters to attend to.

My thanks to all.

Aboriginal burning reference from Cabeza de Vaca, *The Relation of Álvar Núñez Cabeza de Vaca* (first published 1542).

2. A. D. Folweiler, *Fire in the Forests of the United States* (St. Louis, MO: John W. Swift, 1946); Henry A. Wright and Arthur W. Bailey, *Fire Ecology in the United States and Southern Canada* (New York: John Wiley and Sons, 1982).
3. The relevant studies, in sequence: Urban Wildland Interface Division, "Cross Plains, Texas Wildland Fire Case Study" (Texas Forest Service, 2007); David Zane et al., "Surveillance of Mortality During the Texas Panhandle Wildfires (March 2006)" (Texas Department of State Health Services, 2007); Karen Ridenour et al., "Texas 2008 Wildfire Season: Central Branch Significant Wildfire Report" (Texas Forest Service, 2009); Karen Ridenour et al., "The Montague Complex Fire, Montague County: A Case Study" (Texas Forest Service, 2009); Karen Ridenour et al., "1148 Complex Fire, Palo Pinto County, April 9, 2009: A Case Study" (Texas Forest Service, 2009); Rich Gray et al., "Wilderness Ridge Fire, Bastrop County, The Most Destructive Wildfire in Central Texas: A Case Study" (Texas Forest Service, 2009).
4. Source: Texas Forest Service. See also "Preliminary Estimates Show Hundreds of Millions of Trees Killed by 2011 Drought," TFS (December 19, 2011).
5. My use of "persuasion" derives from the example of Marvin Meyers, *The Jacksonian Persuasion: Politics and Belief* (New York: Vintage Books, 1960).
6. Among references for understanding the Texas experience, I found D. W. Meinig, *Imperial Texas: An Interpretive Essay in Cultural Geography* (Austin: University of Texas Press, 1969) and the old classic, Walter Prescott Webb's *The Great Plains* (1931; repr., Lincoln: University of Nebraska Press, 1981) to be among the most useful. Archie P. McDonald, *Texas: A Compact History* (Abilene, TX: State House Press, 2007) was helpful for keeping the chronology straight.
7. For the genesis of the TFS, see David Lane Chapman, "An Administrative History of the Texas Forest Service, 1915–1975" (PhD dissertation,

Texas A&M University, 1981). Unfortunately the study ends two decades before the modern era of fire protection.

8. On its population shifts, see http://osd.texas.gov/Data/TPEPP/Estimates/.
9. On the changing composition of Texas lands, see Neal Wilkins et al., "Fragmented Lands: Changing Land Ownership in Texas," MKT-3443 (Texas A&M Agricultural Communications, 2000).
10. Texas Forest Service, *Cross Plains, Texas Wildland Fire Case Study* (Texas Forest Service, 2007).
11. For programs to assist fire departments, see http://texasforestservice.tamu.edu/FireDepartmentPrograms/.
12. Numbers from Tom Spencer, TFS. Staff included 213 firefighting personnel, and 79 personnel involved in other fire operations.
13. Webb, *Great Plains*, 166. Regrettably (for the cause of easy symbolism), TFS personnel are not called "rangers," as most forestry personnel are. But the analogy is based on actions, not titles.
14. T. L. Mitchell, *Journal of an Expedition into the Interior of Tropical Australia* (London: Longman, Brown, Green, and Longmans, 1848), 306. "Red buffalo" from Courtwright, *Prairie Fire*, 13.
15. For a distilled, popular summary, see Jim Ansley and Charles Hart, "Drivers of Vegetation Change on Texas Rangelands," *AgriLife Extension*, L-5534, Texas A&M AgriLife Communications, February 2012.
16. The literature on brush is as extensive as its topic and has been the subject of numerous conferences. For the particular areas I visited, I found the following publications useful as supplements to field tours by hosts: W. R. Teague et al., "Integrated Grazing and Prescribed Fire Restoration Strategies in a Mesquite Savanna: I. Vegetation Responses," *Rangeland Ecology and Management* 63, no. 3 (2010): 275–85, and R. J. Ansley et al., "Integrated Grazing and Prescribed Fire Restoration Strategies in a Mesquite Savanna: II. Fire Behavior and Mesquite Landscape Cover Responses," *Rangeland Ecology and Management* 63, no. 3 (2010): 286–97; R. James Ansley and G. Allen Rasmussen, "Managing Native Invasive Juniper Species Using Fire," *Weed Technology* 19 (2005): 517–22; Robert K. Lyons et al., "Juniper Biology and Management in Texas," *AgriLife Extension*, B-6074, Texas A&M AgriLife Communications, 2009.
17. Quote from interview with Jim Ansley, March 11, 2012.
18. Statistics from Texas Forest Service.

19. See Big Bend National Park's website at https://www.nps.gov/bibe/learn/management/losdiablos.htm.
20. The literature on central Texas brush is diffused over symposia proceedings, numerous disciplinary journals, and local publications and handouts. For a good sample of the recent science, see Charles A. Taylor Jr. et al., "Long-Term Effects of Fire, Livestock Herbivory Removal, and Weather Variability in Texas Semiarid Savanna," *Rangeland Ecology and Management* 65, no. 1 (2012): 21–30.
21. An excellent overview of early pastoralism is available in Bonney Youngblood, *An Economic Study of a Typical Ranching Area on the Edwards Plateau of Texas*, Texas Agricultural Experiment Station Bulletin No. 297 (Texas A&M, 1922); quote from p. 70.
22. For an excellent overview, see Charles A. Taylor Jr., "Prescribed Burning Cooperatives: Empowering and Equipping Ranchers to Manage Rangelands," *Rangelands*, February 2005, 18–23. Developments can be tracked through the Edwards Plateau Prescribed Burning Association Inc. newsletter, *EPPBA News*.
23. My sources for the Bastrop fire come from a conversation with Greg Creacy and a four-day series of articles in the *Austin American-Statesman*, March 4–7, 2012. Of particular value for reconstructing the dynamics was the feature by Dave Harmon, "Unstoppable: Anatomy of Texas' Most Destructive Wildfire," March 3, updated March 5. The Texas Forest Service has a detailed analysis of the fire, but it remains under internal review and was not available for my examination. An interesting personal account of the fire and its impact, written long after my own text, is Randy Fritz, *Hail of Fire: A Man and His Family Face Natural Disaster* (San Antonio, TX: Trinity University Press, 2015).
24. John Steinbeck, *Travels with Charley in Search of America* (1962; repr. New York: Penguin, 1980), 177.
25. Larry McMurtry, *In a Narrow Grave: Essays on Texas* (New York: Simon and Schuster, 1968), 164.
26. Ibid.
27. In a *New York Times* interview in 1997, McMurtry exclaimed he was "bored to death with the 19th-century West." Yet, speaking from where he grew up, he also remarked, "It's still such a strong landscape for me. I can't escape it in my fiction. I can work away from it, but I always start here. And whatever piece I'm writing about, I'm still describing this same hill." Mark Horowitz, "Larry McMurtry's Dream Job," *New*

York Times Book Review, December 7, 1997, http://www.nytimes.com/books/97/12/07//home/article2.html.

28. The phrase was not originally coined as a call to arms for continental expansion but as a meditation on the power of the American republic to inspire similar revolutions across North America, of which Texas was the first. See Frederick Merk, *Manifest Destiny and Mission in American History* (New York: Knopf, 1963).

PEOPLE OF THE PRAIRIE, PEOPLE OF THE FIRE

1. This piece of fire journalism is the outcome of a too-fleeting visit at the invitation of the Illinois Prescribed Fire Council to speak at their annual meeting in May 2009, spiced with field trips to Kankakee and Nachusa. I would like to thank in particular Fran Harty, Bill Kleiman, and Cody Considine for their roles as hosts and tutors. Anyone familiar with the pyric geography of these sites will appreciate that I add nothing to data or concepts. Instead, I have sought only to establish a different perspective and narrative for their understanding.

   Since the prairie peninsula helps bound themes from the Great Plains, I have included the essay with this survey.

WICHITA MOUNTAINS

1. I want to thank Ralph Godfrey for helping set up a tour of the refuge even though he was not, in the end, able to participate. Special thanks to Richard Baker, Walter Munsterman, and Aaron Roper for interrupting their regular duties and preferred tasks to explain some of the basics to an eager but ill-informed visitor.
2. The best (only) source of fire history data is Michael C. Stambaugh et al., "Fire, Drought, and Human History near the Western Terminus of the Cross Timbers, Wichita Mountains, Oklahoma, USA," *Fire Ecology* 5, no. 2 (2009): 51–65.

THE BLACKENED HILLS

1. Blaine Cook did yeoman work in setting up a panel discussion with the key fire staff at the forest, and then organizing a marvelous field tour. For much that is informed about my essay, he deserves credit—and has my thanks. For their particular comments I would also like to

thank Amy Ham, Jason Virtue, Rochelle Plocek, Frank Carroll, Todd Pechota, and Dave Mertz. I think it fair to say that there are passages that each one—each for his or her own reasons—would disagree with. The hills deserve to move from outlier to core in the national debate.

2. On its lusty pine, see Wayne D. Shepperd and Michael A. Battaglia, *Ecology, Silviculture, and Management of Black Hills Ponderosa Pine*, General Technical Report RMRS-GTR-97, U.S. Forest Service, 2002.
3. The photos have inspired several classic rephotography projects beginning with Donald R. Progulske, *Yellow Ore, Yellow Hair, Yellow Pine*, Bulletin 616, South Dakota Agricultural Experiment Station, 1974, and most recently, Ernest Grafe and Paul Horsted, *Exploring with Custer: The 1874 Black Hills Expedition* (Custer, SD: Golden Valley Press, 2002).
4. See, for example, Peter M. Brown and Carolyn Hull Sieg, "Fire History in Interior Ponderosa Pine Communities of the Black Hills, South Dakota, USA," *International Journal of Wildland Fire* 6, no. 3 (1996): 97–105. Other data come from fire-management staff at Black Hills National Forest.
5. Benjamin Kleinjan, "A Brief History of the Black Hills Forest Fire Protection District," 2009, unpublished report for the South Dakota Department of Agriculture.
6. Statistics from Black Hills National Forest fire-management office.
7. Environmental critiques seem sparse, but a slight introduction is available in Sven G. Froiland, *Natural History of the Black Hills and Badlands* (Sioux Falls, SD: Center for Western Studies, Augustana College, 1999).

EPILOGUE

1. Willa Cather, "Nebraska: The End of the First Cycle," in *O Pioneers!*, ed. Sharon O'Brien, Norton Critical Editions (New York: Norton, 2008): 337.
2. In *Prairie Fire* (pages 188–89) Julie Courtwright turns her conclusion on a written account of an 1867 fire in which, to the observer, a locomotive and prairie fire seemed to race for the future.
3. Quote from Deborah Epstein Popper and Frank J. Popper, "The Great Plains: From Dust to Dust," *Planning*, December 1987, 18.

INDEX

ABOUT THE AUTHOR

Stephen J. Pyne is a historian in the School of Life Sciences at Arizona State University. He is the author of over 30 books, mostly on wildland fire and its history but also dealing with the history of places and exploration, including *The Ice*, *How the Canyon Became Grand*, and *Voyager*. His current effort is directed at a multivolume survey of the American fire scene—*Between Two Fires: A Fire History of Contemporary America* and *To the Last Smoke*, a suite of regional reconnaissances, all published by the University of Arizona Press.